Applied

Inorg

Applied Science Review™

Inorganic Chemistry

Mary Jean Rutherford, MEd, MT(ASCP)SC
Program Director
Medical Technology and Medical Technicians—AS Programs;
Assistant Professor in Medical Technology
College of Nursing and Health Professions
Arkansas State University
State University

Springhouse Corporation
Springhouse, Pennsylvania

Staff

EXECUTIVE DIRECTOR, EDITORIAL
Stanley Loeb

DIRECTOR OF TRADE AND TEXTBOOKS
Minnie B. Rose, RN, BSN, MEd

ART DIRECTOR
John Hubbard

CLINICAL CONSULTANT
Maryann Foley, RN, BSN

EDITORS
Diane Labus, Paula Bakule

COPY EDITOR
Pamela Wingrod

DESIGNERS
Stephanie Peters (associate art director), Matie Patterson

COVER ILLUSTRATION
Scott Thorn Barrows

ILLUSTRATORS
Jacalyn Facciolo, Rhonda Forbes

TYPOGRAPHY
David Kosten (director), Diane Paluba (manager), Elizabeth Bergman, Joyce Rossi Biletz, Phyllis Marron, Robin Mayer, Valerie L. Rosenberger

MANUFACTURING
Deborah Meiris (manager), Anna Brindisi, T.A. Landis

©1993 by Springhouse Corporation, 1111 Bethlehem Pike, P.O. Box 908, Springhouse, PA 19477-0908. All rights reserved. Reproduction in whole or part by any means whatsoever without written permission of the publisher is prohibited by law. Authorization to photocopy items for internal or personal use, or the internal or personal use of specific clients, is granted by Springhouse Corporation for users registered with the Copyright Clearance Center (CCC) Transactional Reporting Service, provided the base fee of $00.00 per copy plus $.75 per page is paid directly to CCC, 27 Congress St., Salem, MA 01970. For those organizations that have been granted a license by CCC, a separate system of payment has been arranged. The fee code for users of the Transactional Reporting Service is 0874344549/93 $00.00 + $.75.
Printed in the United States of America.
ASR2-010792

Library of Congress Cataloging-in-Publication Data
Rutherford, Mary Jean.
   Inorganic chemistry / Mary Jean Rutherford.
     p. cm. — (Applied science review)
   Includes bibliographical references and index.
   1. Chemistry, Inorganic. I. Title. II. Series
   QD151.2.R88 1993   92-18770
   546—dc20   CIP

ISBN 0-87434-454-9

Contents

Advisory Board and Reviewers vi
Acknowledgments and Dedication vii
Preface viii

1. **Overview of Chemistry and Measurement** 1

2. **Matter and Energy** 9

3. **Elements: The Building Blocks of Matter** 16

4. **Electrons and Ions** 19

5. **Chemical Bonds and Formulas** 26

6. **Chemical Reactions and Equilibrium** 35

7. **Solutions** 42

8. **Solids** 47

9. **Liquids** 52

10. **Gases** 61

11. **Radioactivity and Nuclear Reactions** 71

12. **Acids, Bases, and Buffers** 79

13. **Metals and Metalloids** 88

14. **Nonmetals and Noble Gases** 96

Appendices

A: Scientific Notation 103
B: Table of Elements 106
C: Periodic Table of Elements 110
D: Glossary 112

Selected References 117

Index 118

Advisory Board

Leonard V. Crowley, MD
  Pathologist
  Riverside Medical Center
  Minneapolis;
  Visiting Professor
  College of St. Catherine, St. Mary's
    Campus
  Minneapolis;
  Adjunct Professor
  Lakewood Community College
  White Bear Lake, Minn.;
  Clinical Assistant Professor of Laboratory
    Medicine and Pathology
  University of Minnesota Medical School
  Minneapolis

David W. Garrison, PhD
  Associate Professor of Physical Therapy
  College of Allied Health
  University of Oklahoma Health Sciences
    Center
  Oklahoma City

Charlotte A. Johnston, PhD, RRA
  Chairman, Department of Health
    Information Management
  School of Allied Health Sciences
  Medical College of Georgia
  Augusta

Mary Jean Rutherford, MEd, MT(ASCP)SC
  Program Director
  Medical Technology and Medical
    Technicians—AS Programs;
  Assistant Professor in Medical Technology
  Arkansas State University
  College of Nursing and Health
    Professions
  State University

Jay W. Wilborn, CLS, MEd
  Director, MLT-AD Program
  Garland County Community College
  Hot Springs, Ark.

Kenneth Zwolski, RN, MS, MA, EdD
  Associate Professor
  College of New Rochelle
  School of Nursing
  New Rochelle, N.Y.

Reviewers

Joanne U. Cerrato, MT(ASCP), MA
  Department Chairperson and Program
    Director
  Medical Laboratory Technology Program
  Springfield (Mass.) Technical Community
    College

Lester Hardegree, MT(ASCP), MEd
  Medical Technology Program Director
  Armstrong State College
  Savannah, Ga.

Acknowledgments and Dedication

I would like to express my gratitude and appreciation to my colleagues in the Medical Technology programs of Arkansas State University for their support and cooperation during the time spent writing this manuscript. I also would like to thank Maryann Foley of Springhouse Corporation for her infinite patience and perseverance during the entire process. And I would like to thank my family for their wholehearted help and support. This book would not have been possible without all these people.

Lastly, I would appreciate hearing from the students and teachers who use this book. I hope they find the book beneficial—that it facilitates learning, increases knowledge, and provides pleasure in their ultimate understanding of chemistry.

With great love to my mother, who always said, "Yes, you can."

Preface

This book is one in a series designed to help students learn and study scientific concepts and essential information covered in core science subjects. Each book offers a comprehensive overview of a scientific subject as taught at the college or university level and features numerous illustrations and charts to enhance learning and studying. Each chapter includes a list of objectives, a detailed outline covering a course topic, and assorted study activities. A glossary appears at the end of each book; terms that appear in the glossary are highlighted throughout the book in boldface italic type.

Inorganic Chemistry provides conceptual and factual information on the various topics covered in most inorganic chemistry courses and textbooks and focuses on helping students to understand:
• principles of measurement and balancing chemical equations
• the states and composition of matter
• the properties and characteristics of atoms, electrons, ions, molecules, and compounds
• chemical bonding and reactions
• the physical and chemical properties of solids, liquids, and gases
• radioactivity and nuclear reactions
• the characteristic properties of metals, metalloids, and nonmetals.

1

Overview of Chemistry and Measurement

Objectives
After studying this chapter, the reader should be able to:
- Define chemistry.
- List the major branches of chemistry and tell which areas of study they cover.
- Define matter, element, molecule, and compound.
- State the rule for determining the number of significant figures in multiplication and subtraction problems.
- Identify the units employed in the metric, Systeme Internationale (SI), and English systems of measurement.
- Write the formulas for volume, velocity, density, force, pressure, energy, and specific gravity.

I. The Science of Chemistry

A. General information
1. *Chemistry*—the study of the composition and interaction of matter—can be broken down into several fields
2. *Inorganic chemistry*—the focus of this book—provides a basic knowledge of chemistry and the principles related to all forms of matter
    a. Traditionally, inorganic chemistry focused on the study of all compounds except those containing carbon
    b. More recently, the study has been broadened to provide a basic understanding of all the elements, including carbon, as a requisite to learning about the more specialized branches of chemistry
3. *Organic chemistry* is the study of substances that contain carbon, usually in combination with such other elements as hydrogen, nitrogen, and sulfur
4. *Physical chemistry* is the study of how the principles of physics are applied to chemistry, such as how the surrounding temperature affects a chemical reaction
5. *Biochemistry* is the study of substances found within living organisms, such as proteins, fats, and deoxyribonucleic acid (DNA)

B. Particles of matter
1. **Matter** is any substance that has mass (weight) and occupies space; it may exist in any of three forms (solid, liquid, or gas)

2. An ***element*** is a pure substance that cannot be broken down into smaller substances; for this reason, elements are considered the building blocks of all matter
3. Chemists study the composition of elements and how those components react with one another and with the components of other elements
   a. All elements are composed of ***atoms*** — minute particles that possess all of the properties of an element and that cannot be broken down into smaller particles without losing their elemental identity
   b. Atoms combine with other atoms to form molecules and compounds
      (1) A ***molecule*** is two or more atoms joined together as a single unit
      (2) A ***compound*** is a substance formed by the union of two or more elements

II. Chemical Measurements

A. General information
1. Chemistry is an experimental science in which substances are measured and expressed as numbers
2. Chemists commonly measure the temperature, mass, length, volume, density, and velocity of various substances, as well as the time, force, pressure, and energy required to change substances into other forms
3. Because most chemical measurements involve extremely large or extremely small numbers, scientists have devised an easy way of expressing numbers as powers of 10 called ***scientific notation***
4. Most chemists use the metric system, which is based on multiples of 10, to record measurements

B. Principles of measurement
1. When measuring substances, chemists must be careful about the accuracy and precision of the measurements
   a. *Accuracy* is a reflection of how closely the measurement comes to the true, accepted value; it is a reflection of correctness (for example, suppose you weighed a gold nugget on a scale and found it to weigh 23 grams [g]; if the actual weight of the gold nugget is 25 g, your scale or your weighing technique caused you to come up with an inaccurate number)
   b. *Precision* is a reflection of how closely a series of measurements reproduce each other exactly; it also is a reflection of consistency (for example, suppose you weighed that same gold nugget two more times on the same scale and found its weight to be 23.1 g and 22.9 g; because all three of your measurements are close to each other, the weight recorded in the three different measurements is precise but inaccurate because the nugget actually weighs 25 g; your measurements are consistent but incorrect)
2. The precision of a measurement is implied by the number of significant figures used in the measured value

Calculating with Significant Figures

Using a ruler that indicates both centimeters and inches, you can measure the width of a standard sheet of typing paper as 21.5 cm, or 8.5 inches. Suppose you wanted to use these measurements to determine how many centimeters are in 1 inch. And suppose you perform the calculation using an electronic calculator.

Step 1. 8.5 in = 21.5 cm

Step 2. $1 \text{ in} = \dfrac{21.5 \text{ cm}}{8.5}$

Step 3. 1 in = 2.529411765 cm (calculator answer)

No measurement with ordinary rulers is accurate enough to give the calculator answer. Because this is a division problem, the final answer can only have as many significant figures as the initial number with the fewest significant figures.

Step 4. 1 in = 2.5 cm

      a. A *significant figure* refers to the number of meaningful digits in a measured or calculated quantity; the number of significant figures in a measurement includes all the numbers that can be read with confidence plus one last figure that is an estimate
      b. The number of significant figures recorded in a measurement varies according to the expected precision of the measurement
      c. The last digit of any measurement is understood to be uncertain because of the limitations on the precision of the measuring instrument and because the last digit is almost always an estimate
  3. Scientists follow a specific set of rules to determine which numbers qualify as significant figures
      a. Any digit that is not a zero is significant (for example, the number 1459.3 has five significant figures)
      b. Zeros between nonzero digits are significant (for example, the number 1459.03 has six significant figures)
      c. If a number is less than 1, zeros to the left of the first nonzero digit are not significant because these zeros simply indicate the placement of the decimal point (for example, 0.014593 has five significant figures)
      d. If a number is greater than 1, then all the zeros written to the right of the decimal point count as significant figures (for example, 1,4593.00 has seven significant figures)
      e. For numbers that do not contain decimal points, the trailing zeros may or may not be significant (for example, 1,459,300 has an unknown number of significant figures); to indicate which trailing zeros are significant, scientists write these numbers in scientific notation (see section C., Scientific notation, for an explanation)
      f. In multiplication and division, the number of significant figures allowed in the answer is limited by the initial number with the lowest number of significant figures (see *Calculating with Significant Figures*)

4 Overview of Chemistry and Measurement

Adding Significant Figures

Find the sum of the following:

Step 1. 462.20 g
          32.210 g
        + 1.325 g
        ──────────
         495.735 g (intermediate answer)

The final answer may only have as many digits to the right of the decimal point as the initial number with the smallest number of digits to the right of the decimal. The first number in the sum has only two digits to the right of the decimal, whereas the other two numbers have three. Therefore, the final answer may have only two digits to the right of the decimal.

Step 2. 495.74 g (final answer, rounded off according to significant figure rules)

---

      g. In addition and subtraction, the answer may contain only as many digits to the right of the decimal point as the initial number with the lowest number of digits to the right of the decimal point (see *Adding Significant Figures*)

      h. To round a number to the desired number of significant figures, one or more digits to the right of the final significant figure should be dropped

      i. When the first digit dropped is less than 5, the last significant figure should remain unchanged

      j. When the first digit dropped is equal to or greater than 5, the last significant figure should be increased by 1

      k. For calculations involving more than one step, the calculations should be completed without rounding the intermediate results; however, the final answer should be rounded to the appropriate number of significant figures

C. Scientific notation
1. Measurements obtained through experiment are typically extremely large or extremely small; therefore, scientists have devised a mathematical system called *scientific notation* to allow the easy use of extremely large or small numbers
2. All numbers in scientific notation can be expressed in the following form: $N \times 10^n$
   a. "N" is a number between 1 and 10; it always shows the desired number of significant figures
   b. The number "n" is an exponent of 10; it may be a positive or negative number or zero
   c. To find "n" for a number greater than 1, count the number of digits between the first number in the figure and the decimal point; the number of digits between the first figure and the decimal point becomes the exponent of 10 (for example, $14{,}593.0 = 1.45930 \times 10^4$)

Commonly Used Prefixes for Metric Units

The chart below includes some of the most commonly used prefixes in the metric system along with their abbreviations and designated power of 10.

PREFIX	ABBREVIATION	MEANING	POWER OF 10
Mega-	M	One million (1,000,000)	10^6
Kilo-	k	One thousand (1,000)	10^3
Deci-	d	One tenth (1/10)	10^{-1}
Centi-	c	One hundredth (1/100)	10^{-2}
Milli-	m	One thousandth (1/1,000)	10^{-3}
Micro-	μ	One millionth (1/1,000,000)	10^{-6}
Nano-	n	One billionth (1/1,000,000,000)	10^{-9}
Pico-	p	One trillionth (1/1,000,000,000,000)	10^{-12}

    d. To find "n" for a number less than 1, count the number of places that the decimal point must be moved to the right to give a number "N" that is between 1 and 10; this number of places becomes the negative exponent for 10 (for example, $0.0014593 = 1.4593 \times 10^{-3}$; see *Appendix B: Scientific Notation* for examples of how calculations may be carried out using scientific notation)

    e. In scientific notation, the number 1 is 10^0 (ten to the zero power)

D. Units of measurement
1. All numbers require some unit to indicate what is actually measured
2. In chemistry, the **metric system** is the most commonly used system of measurement; other systems with limited applications include the **Systeme Internationale** and the **English system**
3. The *metric system,* which is widely used for scientific measurements, is a comparitively easy system in which units are converted to more usable units based on multiples or divsions of 10; so that converted units can be easily recognized and readily understood, each multiple or subunit of 10 is assigned a separate prefix (see *Commonly Used Prefixes for Metric Units*)
    a. Length is measured in meters
    b. Mass (weight) is measured in kilograms
    c. Time is measured in seconds
    d. Temperature is measured in Celsius degrees (C°)
4. The *Systeme Internationale* (or SI system), a modernized version of the metric system, is the preferred system of measurement for health care because it uses standardized units for all types of measurements; like the metric system, the SI system uses units based on multiples or divisions of 10
    a. Length is measured in meters

6 Overview of Chemistry and Measurement

Metric Conversion Table

Below are some commonly used conversions for mass, length, and volume in the metric system.

MASS	LENGTH	VOLUME
1 gram = 1,000 milligrams	1 meter = 1,000 millimeters	1 liter = 1,000 milliliters or 1,000 cubic centimeters*
1 gram = 100 centigrams	1 meter = 100 centimeters	
1 gram = 10 decigrams	1 meter = 10 decimeters	1 liter = 100 centiliters
1 kilogram = 1,000 grams	1 kilometer = 1,000 meters	1 liter = 10 deciliters
		1 kiloliter = 1,000 liters

*The terms milliliter and cubic centimeter may be used interchangeably (1 mm = 1 cc); some scientists use milliliters when referring to liquid volumes and cubic centimeters when referring to gaseous volumes.

---

    b. **Mass** is measured in kilograms
    c. Time is measured in seconds
    d. Temperature is measured in Kelvin degrees (K°)
  5. The *English system,* also known as the U.S. customary system, is rarely used for scientific purposes because conversion between units of similar measurement is difficult; this system uses various bases, not a single base, for all measurements (for example, 12 inches equal 1 foot; 3 feet or 36 inches equal 1 yard)
    a. Length is measured in feet
    b. Mass is measured in pounds
    c. Time is measured in seconds
    d. Temperature is measured in Fahrenheit degrees (F°)

E. Volume measurements
  1. *Volume*—the amount of space an object occupies—can be expressed according to the following equation: $V = l \times w \times d$, where "V" is volume, "l" is length, "w" is width, and "d" is depth
  2. The unit for volume in the metric system is the liter, which is equal to 1,000 cubic centimeters, just slightly larger than a quart
  3. Because the liter is such a large volume and laboratory experiments typically involve only small quantities of substances, scientists commonly measure volumes in milliliters (see *Metric Conversion Table* for a list of unit conversions within the metric system)

F. Velocity measurements
  1. *Velocity*—the change in distance over time—can be expressed by the following equation: $v = d/t$, where "v" is velocity, "d" is distance, and "t" is time
  2. Velocity measurements are needed to define acceleration, force, and energy
  3. Velocity is expressed in the metric system as meters per second (m/s) or centimeters per second (cm/s)

G. Force measurements
  1. *Force*—the change in velocity of a mass over time—is expressed as follows: $F = m \times a$, where "m" is the mass of an object and "a" is its acceleration (change of velocity over time)

2. In the metric system, force is expressed in units called *dynes;* one dyne is the force required to give a 1-gram mass an acceleration of one centimeter per second
3. Force is expressed in the SI system as newtons (N); one newton is the force required to give a mass of one kilogram an acceleration of one meter per second per second
4. Chemists are concerned primarily with the electrical forces that exist among atoms and molecules

H. **Pressure measurements**
1. *Pressure*—the force applied per unit of area—is expressed by the following equation: P = F/a, where "P" is the pressure, "F" is the force, and "a" is the area over which the force is exerted
2. The metric system expresses pressure in millimeters of mercury (mm Hg); this unit is commonly used to determine atmospheric pressure or the pressure of a gas inside a closed container
3. The SI unit of pressure is the pascal (Pa); 1 pascal equals 1 newton per square meter

I. **Energy measurements**
1. *Energy*—the amount of force applied over a certain distance—is a measure of the capacity to do work or to produce change; it is expressed as follows: E = F × d, where "E" is energy, "F" is applied force, and "d" is the distance over which the force is applied
2. Energy is traditionally measured in the metric system in calories; a calorie is the amount of heat required to raise the temperature of 1 gram of water 1° C; in nutrition, heat is expressed as Calories, where 1 Calorie is equal to 1,000 calories
3. The SI unit of energy is the joule (J); 1 calorie equals 4.184 joules

J. **Density measurements**
1. The *density* of an object—the mass of the object per unit of volume—can be expressed as follows: D = m/V, where "D" is the density, "m" is the mass of the object, and "V" is the volume of the object
2. The metric or SI unit for density is kilogram per cubic meter (kg/m^3), or more conveniently, as gram per cubic centimeter (g/cm^3)

K. **Specific gravity measurements**
1. *Specific gravity* is the ratio of the mass of a given volume of a substance to the mass of an equal volume of water measured at the same temperature; the specific gravity of various body fluids is commonly measured in health care
2. Specific gravity is expressed by the following formula:

$$sg = \frac{d_{(sample)}}{d_{(water)}}$$

where "sg" is the specific gravity, "$d_{(sample)}$" is the density of the sample at a given temperature and "$d_{(water)}$" is the density of water at the same temperature

Comparing Temperatures Among Different Systems

This chart compares the values of some commonly measured temperatures using the metric, English, and SI systems.

TEMPERATURE MEASURED	METRIC SYSTEM (CELSIUS)	ENGLISH SYSTEM (FAHRENHEIT)	SI SYSTEM (KELVIN)
Boiling point of water	100° C	212° F	373° K
Freezing point of water	0° C	32° F	273° K
Absolute zero	−273° C	−459.4° F	0° K
Human body temperature	37° C	98.6° F	300° K

L. Temperature measurements
1. *Temperature* is a measure of heat intensity; qualitatively it describes the amount of heat in a given quantity of matter
2. Temperature can be measured according to three different systems or scales— Fahrenheit, Celsius, or Kelvin (see *Comparing Temperatures Among Different Systems* for examples of some common measurements)
    a. The Fahrenheit scale, which is commonly used in commerce and industry, designates the freezing point of water at 32° F and the boiling point of water at 212° F
    b. The Celsius scale (formerly called the Centigrade scale) is widely used for scientific purposes; this system designates the freezing point of water as 0° C and the boiling point of water as 100° C
    c. The Kelvin scale—the official temperature scale for the SI system—is based on the concept that liquid water, ice, and water vapor can exist together in the absence of air or another substance at only one temperature called the ***triple point of water*** (273.16° K); the lowest point on the scale (0° K) is called ***absolute zero*** because no substance can be cooled below this point
    d. The size of the degree used in the Celsius and Kelvin scales is exactly the same; the difference between the two scales is the assignment of the zero value

---

Study Activities

1. Compare and contrast inorganic and organic chemistry.
2. Define accuracy and precision.
3. Express your height and weight (mass) in scientific notation.
4. List commonly used prefixes in the metric system along with their abbreviations.
5. Explain how to convert kilograms into centigrams, decigrams, and milligrams.
6. Walk across the room and calculate your velocity.
7. Calculate the density of this book.

2

Matter and Energy

Objectives

After studying this chapter, the reader should be able to:
• Identify the three states of matter.
• State the difference between a chemical change and a physical change in matter.
• List the four types of energy.
• Describe the difference between kinetic energy and potential energy.
• Describe the five different types of electromagnetic energy.
• State the law of conservation of matter and energy.

I. Matter

A. General information
1. *Matter* is anything that occupies space and has mass (weight)
   a. Matter can exist in one of three states or forms: solid, liquid, or gas
   b. Each substance of matter retains certain physical or chemical properties that distinguish it from all other substances
2. An *element* is a form of matter that cannot be broken down into simpler substances by ordinary chemical processes; scientists currently classify about 109 substances as elements; elements are the building blocks of all matter
3. An *atom* is the smallest particle of an element that exists and still retains the properties of the element
4. A *molecule* is a group of two or more atoms tightly bound together and functioning as a single unit (for example, two oxygen atoms joined together form a molecule of atmospheric oxygen gas [O_2]; two hydrogen atoms joined with an oxygen atom form a molecule of water [H_2O])
5. A molecule may consist of atoms of two or more different elements; such a molecule has properties different from those of the original elements (for example, the metal sodium [Na] and the gas chlorine [Cl] react together to form table salt [NaCl])
6. A **pure substance** is any form of matter that has a definite, constant composition and distinct properties; it may be either an element or a compound
7. A *compound* is a pure substance produced by the chemical combination of two or more elements
   a. It can be broken down into two or more simpler substances by ordinary chemical means; ultimately, all compounds break down to elements

b. The characteristics of a compound may be totally different from the characteristics of its component elements; for example water—a liquid at room temperature—is a compound of hydrogen and oxygen, both of which are gases ordinarily
8. A *mixture* is a combination of two or more substances (either elements or compounds, or both) in any proportion; each substance in the mixture retains its original identity and, in many cases, can be separated from the other substances by physical means (for example, butter consists of oil and solid materials, which separate when heated; the solids will float on top of the oil)
9. A mixture may be homogeneous or heterogeneous
    a. A *homogeneous mixture* is one in which all component substances have the same uniform characteristics throughout the mixture; for example, sugar dissolved in water is homogenous because the sugar water has the same uniform appearance and composition throughout; a homogeneous mixture in which one substance is dissolved in another is also called a *solution*
    b. A *heterogeneous mixture* is one in which the component substances do not have the same uniform characteristics throughout the mixture; for example, chunky peanut butter is a heterogeneous mixture in which the peanut chunks are visibly different from the peanut butter and the peanut chunks are not distributed evenly throughout the entire mixture; one sample may contain 10 peanut chunks, whereas a second sample of the same size may contain 25 peanut chunks

B. States of matter
1. In nature, matter occurs in only three different forms or states: solid, liquid, or gas
    a. A *solid* holds its shape and volume even when not in a container; the molecules of a solid are tightly compacted and move only slightly
    b. A *liquid* assumes the shape of its container; the molecules of a liquid are in constant motion and do not have the fixed arrangement found in solids
    c. A *gas* has no shape, diffuses readily, and assumes the full-volume shape of any closed container; gas molecules are widely distributed and can move in any direction
2. Scientists typically describe each state of matter in terms of four characteristics: the distance between molecules, the forces of attraction between molecules, the motion of the molecules, and the arrangement or orderliness of the molecules (see *Comparing States of Matter*)
3. The same matter can change from one state to another, depending on the temperature and pressure of the surrounding environment; for example, water may be a liquid, a solid (ice), or a gas (steam)
4. Matter undergoes a *physical change* (*change of state*) when it physically changes from one form to another, such as from a solid to a liquid; a change of state does not alter the basic chemical composition of matter
    a. An *endothermic change of state* is one that absorbs energy during the physical change; for example, when solid water (ice) changes to liquid water, energy is absorbed

Comparing States of Matter

Here is a comparison of the characteristics by which scientists distinguish among solids, liquids, and gases at the molecular level.

CHARACTERISTIC	SOLID	LIQUID	GAS
Distance between molecules	Molecules tightly compacted	Molecules close together	Molecules far apart
Motion of molecules	Little movement; lack of fluidness	Flows with varying ease depending on the type of liquid	Flows easily
Arrangement of molecules	Has a definite volume and shape that is virtually incompressible; high density	Has a definite volume, but assumes the shape of its container; slightly compressible; high density	Assumes the volume and shape of its container; very compressible; low density

      b. An *exothermic change of state* releases energy during the physical change; for example, when gaseous water (water vapor or steam) condenses on a cold window to form a film of liquid water, energy is released

      c. **Sublimation** is a change of state in which a substance passes from the solid state directly to the gaseous state at the same temperature; such a process occurs when dry ice (solid carbon dioxide) changes directly to carbon dioxide gas

   5. Matter undergoes a **chemical change** when the basic chemical composition of matter changes to form a different substance; for example, burning wood changes the wood to chemically new substances (ash, charcoal, and gases)

C. Properties of matter

1. Each substance of matter has certain characteristics (properties) that distinguish it from all other substances, including color, taste, weight, and temperature
2. All properties of matter are considered extensive or intensive, depending on the amount of matter observed
   a. *Extensive properties,* such as weight and volume, vary with the amount of matter being observed
   b. *Intensive properties,* such as temperature and density, do not vary with the amount of matter being observed
3. Properties can be either physical or chemical
   a. **Physical properties** are those that can be observed or measured without changing the chemical composition of the substance (for example, the melting point of water can be determined by placing a thermometer next to an ice cube and observing the temperature reading as the ice melts; although the water changes from solid to liquid form, its chemical composition remains the same); the most commonly described physical properties include color, hardness, solubility, density, melting point, and boiling point
      (1) Most substances display a distinctive color

(2) Hardness refers to the strength of the substance; it may range from very hard (as with diamonds) to soft (as with metallic aluminum)
(3) Solubility indicates the maximum amount of a substance that can be dissolved in a given quantity of water at a specific temperature
(4) Density is the mass of a substance divided by its volume
(5) The **melting point** (also called the **freezing point**) is the temperature at which a substance in its solid state changes into its liquid state or vice versa; at this point, the solid and liquid phases of a substance are in equilibrium
(6) The **boiling point** is the temperature at which a substance in the liquid state forms gaseous bubbles that rise to the surface of the liquid and evaporate; at this point, the vapor pressure of a liquid is equal to the external atmospheric pressure; the boiling point of a substance changes with the surrounding atmospheric pressure

b. **Chemical properties** are those exhibited when matter changes its chemical composition
  (1) Matter can change its chemical composition by a chemical reaction with another substance; for example, iron reacts with oxygen and water to form a new substance—ferric oxide (rust)
  (2) Matter also can change its chemical composition by decomposing (breaking apart into its original elements)
  (3) Each substance has an infinite number of chemical properties, depending on the surrounding physical conditions and the number of substances available with which it can react; commonly described chemical properties include those exhibited when a substance is exposed to air, an acid, or heat

II. Energy

A. General information
1. *Energy* is the capacity to do work
2. **Work** is energy released or absorbed by a system through mechanical means (for example, when gasoline is burned in a car engine, it produces energy plus gaseous carbon dioxide and water; through the mechanics of the engine, the energy released by burning gasoline performs the work of moving the car from one point to another; work is simply the visible result of energy being transferred from one part of a system to another)
3. In nature, energy appears in different forms but is generally classified into four major types: kinetic energy, potential energy, heat energy, and electromagnetic energy

B. Kinetic energy
1. **Kinetic energy** is the energy that an object or particle possesses as a result of its mass and motion; in nature, atoms and molecules are in constant motion and therefore possess kinetic energy
2. Scientists usually measure the kinetic energy of a substance with a thermometer and record it as heat; the higher the temperature, the greater the kinetic energy of atoms or molecules in the substance

Comparing Electromagnetic Energy Wavelengths and Frequencies

TYPE OF ENERGY	WAVELENGTH (CM)	FREQUENCY (CYCLES/SECOND)
X-rays	1×10^{-8}	3×10^{18}
Ultraviolet light	2×10^{-5}	1.5×10^{15}
Visible light	5×10^{-5}	0.6×10^{15}
Infrared radiation	1×10^{-3}	3×10^{13}
Radar or microwaves	1	3×10^{10}
Radiowaves	3×10^{5}	1×10^{5}

3. The amount of kinetic energy in a particle or object depends on both its mass (weight) and velocity (speed); for example a bowling ball traveling down a bowling lane at 30 miles per hour has greater kinetic energy than a tennis ball traveling at the same speed because of its larger mass
4. The kinetic energy for any object is represented by the formula: kinetic energy = mv^2, where "m" is the mass of the object and "v" is its velocity

C. Potential energy
  1. In nature, every object or particle is either attracted to or repulsed by objects that surround it; these forces of attraction and repulsion exist even when we cannot observe them easily; for example, the earth attracts objects to itself by the force of gravity; the positive pole of a magnet repels the positive pole of another magnet through the force of magnetic repulsion
  2. *Potential energy* is stored energy associated with the attraction or repulsion of objects or particles; any collection of particles or objects possesses potential energy
  3. Because potential energy is stored energy, it cannot perform work until is is converted to another form of energy, such as kinetic energy or heat energy; for example, the human body converts the potential energy stored in food to other forms of energy, such as heat

D. Heat energy
  1. *Heat energy* is energy transferred from one place to another as a result of a difference in temperature; scientists measure heat transfer in units called calories
  2. A *calorie* is the amount of heat energy required to raise the temperature of one gram of water one degree Celsius
  3. In human nutrition, the term Calorie (with a capital C) really refers to a *kilocalorie,* or 1,000 calories

E. Electromagnetic energy
  1. *Electromagnetic energy* (also called electromagnetic radiation) is energy that travels through space in regularly spaced waves, much as water travels across the ocean in waves (see Comparing Electromagnetic Energy Wavelengths and Frequencies)

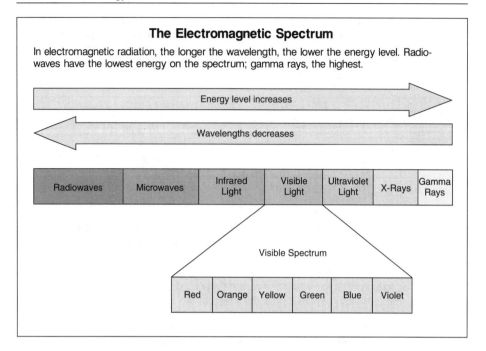

2. Scientists describe electromagnetic energy by noting the wavelength and the frequency of the waves
   a. The *wavelength* of an electromagnetic radiation is the distance between two peaks of the waves; wavelength usually is measured in normal linear units (such as meters or centimeters)
   b. The *frequency* of an electromagnetic radiation is the number of waves that pass by a given point in a given time; frequency usually is measured in cycles per unit of time (for example, cycles per second)
3. In the electromagnetic spectrum, each wavelength corresponds to a specific level of energy; longer wavelengths correspond to lower energy levels, whereas shorter wavelengths correspond to higher energy levels (see *The Electromagnetic Spectrum*)
   a. Radiowaves followed by microwaves have the longest wavelengths and the lowest energy levels; radiowaves include wavelengths used for television; microwaves are used for cooking
   b. Infrared light, the next component of the electromagnetic spectrum, is an invisible light that produces heat; it aids in the identification of substances or compounds from their molecular structure and has special applications in chemistry and health care
   c. Visible light, most commonly observed as white light, is actually composed of the whole spectrum of colors that appear in a rainbow
      (1) Visible light is only a very small portion of the electromagnetic spectrum
      (2) Because certain chemical reactions produce changes in visible light, scientists use visible light as a way of measuring whether or not certain chemical reactions have occurred

(3) Health care scientists use changes in visible light to aid in the chemical analysis of body fluids
- d. Ultraviolet radiation or light has shorter wavelengths and therefore higher energy than visible light; ultraviolet radiation is invisible, but it also has a critical role in the chemical analysis of body fluids
- e. X-rays and gamma rays have the shortest wavelengths and greatest energy levels; both have significant medical applications

F. **Conservation of matter and energy**
1. Scientists have observed that the total amount of matter and energy in the universe remains fixed; this fundamental principle is called *law of conservation of matter and energy*
2. According to the law of conservation of matter and energy, matter may be converted to energy—in which case, mass is lost and energy is created, but the total combined amount of matter and energy in the universe remains the same
3. Energy can be changed from one form to another; it is rarely gained or lost

Study Activities

1. Give two examples of homogeneous mixtures and two examples of heterogeneous mixtures.
2. List four characteristics used to describe each state of matter.
3. Roll a golf ball across the room and determine its kinetic energy.
4. Calculate the calories required to bring one liter of water to the boiling point.
5. Explain how wavelength and frequency are measured.

3

Elements: The Building Blocks of Matter

Objectives
After studying this chapter, the reader should be able to:
- Identify and describe the principal parts of an atom.
- Define isotope, atomic number, mass number, atomic mass, and relative mass.
- Identify the three characteristic numbers used to describe an element.

I. Elements

A. General information
1. Elements are the simplest substances or kinds of matter
2. Each element consists of atoms that are chemically identical to each other and different from the atoms of every other element
3. Of the 109 elements scientists have identified, 85 occur in natural sources within the earth; the remaining elements have only been synthesized in the laboratory; because synthetic elements are unstable, they do not exist in nature
4. Scientists represent element names with a symbol called the ***atomic symbol***
    a. The symbol is usually the capital form of the first letter of the name (for example, the element carbon is represented by the capital letter "C")
    b. In some cases, the first two letters of the element's name are used; the first letter is a capital letter, and the second letter is lowercase (for example, the element calcium is represented by the letters "Ca")
    c. The symbol for some elements is derived from the Latin name for the element; for example, sodium is "Na" (natrium); and, silver is "Ag" (argentum)

B. Elemental components
1. The basic component of all elements is the *atom*
2. All the atoms of the same element have identical properties that are different from the properties of the atoms found in other elements
3. All atoms are electrically neutral because they contain a positively charged nucleus and a negatively charged cloud of electrons surrounding the nucleus; the positive and negative charges balance each other
4. Every atom consists of several types of particles, the most important of which are protons, neutrons, and electrons

a. A ***proton*** is a positively charged particle found in the nucleus of an atom; it has a relative mass of 1 atomic mass unit (amu)
b. A ***neutron*** is a neutral particle (it has no charge) found in the nucleus of an atom; it has a relative mass of 1 amu
c. An ***electron*** is a negatively charged particle that orbits around the nucleus; the number of electrons orbiting the nucleus equals the number of protons in the nucleus; electrons have a relative mass of $1/1,837$ amu

II. Identifying Elements

A. General information
1. Scientists have devised a system of identifying elements based on the mass (weight) and number of particles within the atoms of elements
2. All atoms of a given element have an idiosyncratic number of protons within the nucleus; this number (the atomic number) is indicated as a subscript to the left of the elemental symbol, such as $_ZX$
3. An element may have several characteristic mass numbers, each signifying a different isotope of that element; the mass number for each isotope is indicated as a superscript to the left of the symbol, such as ^AX
4. Each element has one characteristic atomic weight based on the average weight of all the isotopes for that element; this number is expressed in terms of units, specifically ***atomic mass units*** (amu)

B. Atomic number
1. The ***atomic number*** is the number of protons in the nucleus of any atom of an element; it always is represented as a whole number (for example, the element carbon possesses 6 protons in the nucleus; therefore, the atomic number is 6)
2. Because the number of electrons in a given atom always equals the number of protons, the atomic number also indicates the number of electrons in the atom

C. Mass number
1. The ***mass number*** of an atom is equal to the number of protons plus the number of neutrons in the nucleus of the atom (for example, most naturally occurring carbon atoms have 6 protons and 6 neutrons; therefore, the mass number for carbon is 12)
2. Because some atoms of the same element contain more neutrons than those of the "average" atom for that element, a given element may have more than one mass number (for example, some carbon atoms have seven neutrons instead of the usual six, yeilding a mass number of 13; still others have eight neutrons, yeilding a mass number of 14)
3. Atoms of the same element with the same atomic number but different mass numbers are called ***isotopes*** or ***nuclides;*** elements naturally occur as mixtures of isotopes (see *Carbon Isotopes,* page 18, for a breakdown of the composition and abundance of three naturally occurring carbon isotopes)

Carbon Isotopes

This chart indicates the percentage of atoms that normally exist as three of the most common isotopes of the element carbon, along with the isotopes' characteristic number of nuclear particles, atomic number, and mass number.

ISOTOPE CHARACTERISTIC	CARBON-12	CARBON-13	CARBON-14
Percentage	98.89%	1.10%	0.01%
Number of protons	6	6	6
Number of neutrons	6	7	8
Atomic number	6	6	6
Mass number	12	13	14

4. The generic symbol of an isotope includes the symbol for the element, the mass number written to the upper left of the symbol, and the atomic number written to the lower left of the symbol, such as $^A_Z X$

D. Atomic mass
1. Each element has a characteristic **atomic mass** (also called **atomic weight**), a weighted average of all the different isotopes of that element
2. Scientists cannot measure the mass of an individual atom; however, they can measure the relative mass of the atoms of different elements to determine the atomic mass
   a. *Relative mass* is the comparative weight of an element in relation to all of the other elements in a given compound (for example, in a water molecule [H_2O], there are two hydrogen atoms for every oxygen atom; scientists know that for every 16 parts by weight of oxygen, there are two parts by weight of hydrogen; therefore, if 16 units is used as the standard for oxygen, hydrogen would have a relative mass of 1 unit)
   b. Until 1961, scientists determined the relative mass of all known elements based on the arbitrary assignment of 16 units for oxygen; however, because of the need for more precise measurements, they now use the weight of the carbon-12 isotope, which is 12.011 amu (for convenience, this number has been rounded to 12 amu)
3. The current atomic mass scale, which is based on the assignment of 12 amu for the mass of the carbon-12 isotope, ranges from 1 amu (for hydrogen, the lightest element) to over 260 amu (for the heaviest elements)

Study Activities

1. Describe protons, neutrons, and electrons.
2. Draw a picture of an atom showing general placements of the protons, neutrons, and electrons.
3. Explain the difference between atomic number and the mass number.

4

Electrons and Ions

Objectives

After studying this chapter, the reader should be able to:
- Describe the Bohr model and calculate the maximum number of electrons possible in energy levels 1 through 7.
- Discuss the basis for the arrangement of elements in the periodic table.
- State the periodic law.
- List the chemical groups identified in the periodic table.
- Explain the formation of anions and cations.

I. Electron Configuration

A. General information
1. Electrons are negatively charged particles that revolve around the nucleus of an atom
2. Scientists have learned that electrons do not circle the nucleus randomly but rather follow distinct pathways
3. According to the Bohr model—developed by Niels Bohr (1885-1962) in 1913—electrons revolve around the nucleus in orbits similar to the way the planets orbit the sun
4. Over time, scientists have modified the Bohr model to reflect their current understanding of atomic particles; however, the basic model (which applied only to hydrogen atoms) still serves as a useful tool for understanding atomic structure and electron configuration

B. Assumptions of Bohr's model
1. Electrons travel around the nucleus of an atom in spherical orbits; these orbits are referred to as *energy levels* or *shells*
   a. A shell contains electrons of approximately the same energy level traveling at about the same distance from the nucleus
   b. Shells are arranged concentrically around the nucleus and are designated by a letter (K, L, M, N, O...) or a value of n ($n = 1$, $n = 2$, $n = 3$, $n = 4$, $n = 5$...)
   c. The maximum number of shells or energy levels in naturally occurring atoms is 7; thus the 7th shell is called the Q ($n = 7$) shell

2. Each shell corresponds to a certain energy value that belongs to the electrons in that orbit; the greater the distance the electrons are from the nucleus, the higher the amount of energy in the shell
   a. The shell closest to the nucleus—the K ($n = 1$) shell—contains electrons with the lowest amount of energy
   b. The shell farthest from the nucleus—the Q ($n = 7$) shell—contains electrons with the highest amount of energy
3. The farther a shell is from the nucleus, the bigger the shell's radius and the more electrons it can hold
4. An electron cannot occupy a space between shells unless it is in the process of jumping from one shell to another; it can jump to another shell only by absorbing or losing energy
   a. An electron must absorb energy to jump to a shell with a higher energy level; such an electron is said to be in an *excited state*
   b. An electron must lose energy to jump to a shell with a lower energy level; such an electron is said to be in a *ground state*
5. The difference in energy when an electron moves from an excited state to a ground state is a specific amount (or **quantum**) of energy; this energy amount, which is emitted and measured as light, is characteristic of each atom
6. Each of an atom's 7 energy shells may contain only a maximum number of electrons; this maximum number is determined by the formula $2n^2$, where n equals the value of the energy level

C. Electron configurations
1. The arrangement, or configuration, of electrons in an atom is determined not only by the number of shells but also by the manner in which electrons fill those shells
2. All of an atom's shells except the lowest energy shell—the K ($n = 1$) shell—contain a set of closely spaced energy levels called *subshells*, which are designated by lower-case letters *(s, p, d, f, g...)*
   a. The number of subshells in a given shell equals the n value of the shell (for instance, the $n = 2$ shell has 2 subshells; the $n = 3$ shell, 3 subshells; the $n = 4$ shell, 4 subshells; and so on)
   b. Scientists refer to specific subshells by identifying the shell number with the subshell letter (for example, all of the subshells in the $n = 3$ shell are written as 3s, 3p, and 3d; those of shell $n = 4$ as 4s, 4p, 4d, and 4f)
   c. The maximum number of electrons in each subshell depends on the type of subshell and not on the shell number (see *Maximum Electrons in Shells and Subshells*)
      (1) Subshell *s* can accommodate 2 electrons
      (2) Subshell *p* can accommodate 6 electrons
      (3) Subshell *d* can accommodate 10 electrons
      (4) Subshell *f* can accommodate 14 electrons
   d. The total maximum number of electrons in the combined subshells equals the maximum number of electrons allowed in that particular shell as determined by the formula $2n^2$, where n equals the value of the shell's energy level

Maximum Electrons in Shells and Subshells

This chart shows the maximum number of electrons that each of the subshells in the 4 lowest energy shells can accommodate. Note that the last column represents the total number of allowable electrons in each shell as determined by the formula $2n^2$.

SHELL	NUMBER OF SUBSHELLS IN SHELL	MAXIMUM NUMBER OF ELECTRONS IN EACH SUBSHELL				TOTAL NUMBER OF ELECTRONS IN SHELL ($2n^2$)
		s	p	d	f	
K ($n = 1$)	1	2				2
L ($n = 2$)	2	2	6			8
M ($n = 3$)	3	2	6	10		18
N ($n = 4$)	4	2	6	10	14	32

       e. Each subshell is comprised of a set of *orbitals;* each orbital has a different size, shape, and spatial orientation and contains electrons of relatively the same energy level
   3. Electrons fill shells sequentially, from the lowest possible energy level to the highest possible energy level; this is known as the **aufbau principle**
      a. The aufbau principle is based on the notion of progressive building (*aufbau* is German for "building up"); just as protons are added one by one to the nucleus of an atom to build up the element, electrons are similarly added to orbitals to fill subshells and shells
      b. During this buildup, some shells or subshells (or both) may be incompletely filled (see *The Aufbau Principle,* page 22, for a diagram of how electrons fill subshells)
      c. Electron configurations are designated by indicating the shell number, followed by the subshell letter, and then the number of electrons in the subshell (expressed as a superscript to the right of the subshell letter), such as $4p^3$
   4. Scientists group together elements with the same types of partially filled subshells; the **periodic table** is a device for showing which elements belong to each group

II. The Periodic Table

A. General information
   1. The *periodic table* is an arrangement of the elements according to periodicity or repeating nature of the chemical properties of the elements (a copy of the periodic table appears in the Appendices)
      a. Elements with similar chemical properties are grouped together
      b. The elements within a group have relatively similar electron configurations
   2. Elements are arranged according to their increasing atomic number
   3. Each element appears in a box that contains specific information about the element, including the element's symbol, atomic number, and atomic weight

22 Electrons and Ions

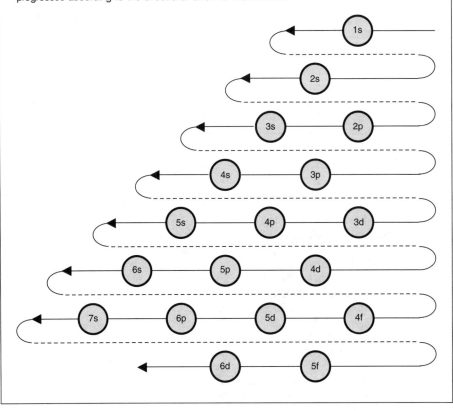

4. The periodic table organizes the elements in both vertical and horizontal classifications based on the types of elements and their physical and chemical properties (see *Periodic Classification of Elements* for a schematic view of how the elements are classified according to the periodic table)
5. **Periodic law** states that when the elements are arranged according to increasing atomic number, their chemical and physical properties show periodic similarities—that is, similarities in properties recur at periodic intervals of atomic number
   a. Electronegativity and ionization energy increase horizontally across the periodic table and decrease vertically down the periodic table
      (1) **Electronegativity** is a measure of how strongly an atom pulls shared electrons toward its nucleus
      (2) **Ionization energy** is the energy required to remove an electron from an atom

Periodic Classification of Elements

In this diagram, each vertical or horizontal bar represents a series of elements with similar chemical characteristics or properties. The bars are arranged according to where the elements appear on the periodic table.

[Diagram of the periodic table showing: Representative elements (IA, IIA, IIIA, IVA, VA, VIA, VIIA, 0); Alkali metals; Alkaline earth metals; Transition elements (IIIB, IVB, VB, VIB, VIIB, VIII, IB, IIB); Halogens; Noble gases; Inner-transition elements (1. Lanthanides, 2. Actinides); Periods 1–7]

    b. As atomic size decreases, ionization energy increases and electronegativity increases; as atomic size increases, ionization energy decreases and electronegativity decreases

B. Periods
1. Each horizontal row in the periodic table is called a *period*
2. Periods are numbered using Arabic numbers from 1 to 7, corresponding to the 7 energy levels (or shells) of an atom that can contain electrons; for example, any element assigned to period 4 will have its outermost electrons assigned to the 4th energy level

C. Groups
1. Each vertical column of the table is called a *group* or *chemical family*

2. The elements in each group or family demonstrate similar chemical behavior as determined by the number of electrons in the outermost energy level; these outermost electrons are called **valence electrons**
3. The vertical columns labeled with Roman numerals I through VII and 0 are called **representative elements;** these elements (also called *main group elements)* are characterized by the progressive addition of electrons to the *s* and *p* subshells
4. Some groups are known by common names
    a. Elements in group I are called **alkali metals**
    b. Elements in group II are called **alkaline earth metals**
    c. Elements in group VII are called **halogens**
    d. Elements in group 0 are called **noble gases** (also called **inert gases**)

D. **Transition elements**
1. Three rows of 10 elements in the middle of the table are called **transition elements**
2. These elements are not included in a specific group designation
3. They are characterized by partially filled *d* or *f* subshells and the addition of electrons to the next-to-outermost energy level rather than the outermost energy level

E. **Inner-transition elements**
1. Two rows of 14 elements at the bottom of the table are termed **inner-transition elements,** characterized by the progressive addition of electrons to the *f* subshell
2. Each of these rows is named for the element it follows in the main body of the periodic table
    a. The *lanthanide series* begins right after the element lanthanum; the series starts with cerium (atomic number 58) and extends to lutetium (atomic number 71)
    b. The *actinide series* begins right after the element actinium; the series starts with thorium (atomic number 90) and extends to lawrencium (atomic number 103)
3. These elements also have similar properties because the addition of electrons occurs below two identical outer-shell electron configurations

F. **Classification of elements by properties**
1. Scientists classify elements by their physical and chemical properties; when classified this way, elements are described as either *metals* or *nonmetals*
2. Metals generally appear on the left and in the center of the periodic table
    a. These elements are generally shiny, dense, malleable, and ductile
    b. Metals have high melting points, are usually solid at room temperature, and conduct heat and electricity more readily than nonmetals
3. Nonmetals appear on the right side of the periodic table
    a. These elements usually have a low density and are brittle when solid
    b. Nonmetals usually are poor conductors of heat and electricity
4. At room temperature, 11 of the known elements on the periodic table are gases, one is a liquid, and the rest are quite brittle solids

5. Some elements called *metalloids* or *semimetalloids* demonstrate some of the properties of both metals and nonmetals; most of these elements appear between metals and nonmetals on the periodic table
6. Hydrogen does not fit into any of the categories described; its physical properties are like those of a nonmetal, but its chemical properties are somewhat like a metal

III. Ions

A. General information
1. In a normal atom, all electrons are found in the lowest possible energy levels (also called the **ground state**)
2. An atom may lose or gain electrons to form a charged atom called an **ion**
3. An ion can have a positive or negative charge depending on whether it gains or loses an electron
   a. A positively charged ion **(cation)** forms whenever an atom loses an electron; its name derives from the fact that the ion moves toward a negatively charged electrical pole called a cathode
   b. A negatively charged ion **(anion)** forms whenever an atom gains an electron; its name derives from the fact that the ion moves toward a positively charged electrical pole called an anode

B. Ion formation
1. If an atom absorbs energy (heat or light) from an external source, it may cause some electrons to jump to higher energy levels; such atoms are said to be in an excited state and are unstable
2. If sufficient energy is absorbed, the atom may lose one of its electrons, leaving the atom with one less electron than its proton; such an atom will no longer have a zero electrical charge but will assume a positive charge
3. Some atoms have incompletely filled high-energy orbits in their outer shells; these atoms will strongly attract weakly held electrons of other atoms to attain greater stability
4. When an atom gains an electron, it assumes a negative electrical charge

Study Activities

1. Draw a Bohr model of an atom showing all seven energy levels and their corresponding subshells.
2. On the same drawing, indicate the maximum number of electrons that can be accommodated in each subshell.
3. Define electronegativity and ionization energy.
4. Using the periodic table, identify three halogens and three alkali metals.
5. Explain the difference between metals and nonmetals.
6. Explain how ions are formed.
7. Define cation and anion and discuss their derivation.

5

Chemical Bonds and Formulas

Objectives

After studying this chapter, the reader should be able to:
- List the three principal types of chemical bonding.
- Compare groups in the periodic table in terms of their ability to form anions and cations.
- Define an ionic compound, and describe how it is formed.
- Identify three polyatomic ions, and describe how they behave chemically.
- Describe how a covalent bond is formed.
- Differentiate polar bonds from nonpolar covalent bonds.
- Describe the type of bonding that characterizes elemental metals, such as silver.
- Name four types of chemical formulas.
- Describe how to determine the formula weight of a chemical compound.

I. Chemical Bonding

A. General information
1. Atoms and ions (atoms with an electric charge) bind together in specific ways to form *compounds*
2. A *molecule* is a compound made of two or more atoms tightly bound together and functioning as a single unit
3. The force that holds atoms or ions together to form a compound is called a **chemical bond;** through chemical bonding, elements may combine to form an infinite number of compounds
4. Three principal types of chemical bonds are ionic bonds, covalent bonds, and metallic bonds

B. The octet rule
1. In the early 1900s, scientists observed that the elements in Groups I through VII of the periodic table form chemical bonds by losing, gaining, or sharing electrons in order to fill their outermost shell of electrons
2. The eight electrons in the outermost atomic shell are called *valence electrons;* atoms with all eight valence electrons are considered most stable
3. The tendency of atoms to form bonds in order to attain an outer octet (eight valence electrons) is called the **octet rule**

a. The octet rule predicts which compounds will be formed from reactions between atoms in Groups I through VII; it works best in predicting the results of reactions between atoms with atomic numbers 1 through 22
   b. Scientists have noted many exceptions to the octet rule, particularly among the heavier elements

II. Ionic Bonds

A. General information
1. In nature, charged objects interact with each other in characteristic ways; objects with like charges (that is, both objects are either positive or negative) repel each other; objects with opposite charges (that is, one object is positive and the other is negative) attract each other
2. An *ionic bond* is a chemical bond that forms when a negatively charged ion attracts and binds to a positively charged ion; this type of bond occurs most commonly when a metallic element reacts with a nonmetallic element
3. Another way to picture an ionic bond is to imagine one atom transferring one or more of its electrons to another atom; for example in the compound sodium chloride, the sodium atom transfers one of its electrons to the chlorine atom
   a. The sodium atom becomes a positive ion because it has lost an electron
   b. The chlorine atom becomes a negative ion because it has gained an electron
   c. The two atoms form an ionic bond based on the attraction between the negatively charged chlorine and the positively charged sodium
4. The attraction between oppositely charged ions, such as a positively charged sodium ion and a negatively charged chlorine ion, is called *electrostatic attraction*
5. Compounds produced through ionic bonding are called *ionic compounds*
   a. Ionic compounds usually are pictured as a three-dimensional array of positive and negative ions, called a *crystalline lattice* (see Sodium Chloride Crystal, page 28, for an illustration)
   b. In this lattice, each positive ion is completely surrounded by negative ions and each negative ion is completely surrounded by positive ions

B. Properties of ionic compounds
1. Ionic compounds usually are crystalline solids at room temperature
2. They tend to have high melting points
3. They dissolve readily in water, freeing the original component ions, each of which becomes surrounded by water molecules
   a. For example, the dissolution of sodium chloride in water allows the positively charged sodium ions and the negatively charged chloride ions to separate and move freely among the water molecules
   b. The solution formed when an ionic compound dissolves in water has an equal number of positive and negative ions
   c. Because of these free ions, water solutions of ionic compounds conduct electricity easily

28 Chemical Bonds and Formulas

Sodium Chloride Crystal
This illustration shows how sodium (Na^+) and chlorine (Cl^-) are arranged in a crystalline lattice to form sodium chloride (NaCl).

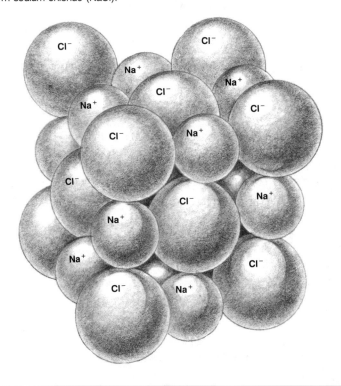

C. Ion formation and the periodic table
1. The periodic table arranges elements in order of how easily they form cations or anions
2. The periodic table also arranges elements in order of their *electronegativity*, or their ability to attract and hold extra electrons; generally, elements on the left side of the table have low electronegativity and elements on the right side of the table have high electronegativity
3. Elements in Group IA (those with low electronegativity) form cations most easily; elements in Group VIIA (those with high electronegativity) rarely form cations
    a. The elements most likely to form cations in ionic compounds are the alkali metals (Group IA) and the alkaline earth metals (Group IIA)
    b. With the exception of hydrogen, all the elements that easily form cations are classified as metals
    c. Although hydrogen is not a metal, it assumes a positive charge when it loses an electron

4. Elements in Group VIIA (those with high electronegativity) form anions most easily; elements in Group IA (those with low electronegativity) rarely form anions
      a. The elements most likely to form anions are the halogens (Group VIIA) and the elements in group VIA
      b. The elements having the highest electronegativity are fluorine, oxygen, chlorine, and bromine

III. Covalent Bonds

A. General information
   1. A **covalent bond** is a chemical bond that forms when two atoms share electrons in mutually held molecular orbitals
      a. The two positively charged atomic nuclei draw closely to each other because they simultaneously attract the same electrons
      b. Compounds produced solely through covalent bonds are called *covalent compounds*
   2. Covalent bonds generally form between atoms of nonmetallic elements; for example, carbon and oxygen join through covalent bonds to form carbon dioxide, and nitrogen and hydrogen join through covalent bonds to form ammonia
   3. When atoms join through covalent bonds, they form a single, electrically neutral molecule
   4. Compounds formed solely of these neutral molecules are called *covalent molecular compounds;* examples of covalent molecular compounds include water and carbon dioxide
   5. Other types of covalent compounds are not composed of discrete, individual molecules; the atoms of these covalent compounds share electrons through extensive three-dimenional networks; silicon dioxide (sand) and diamond are both examples of this type of covalent bonding
   6. Atoms of any group on the periodic table will tend to form covalent bonds with other atoms that have about the same degree of electronegativity
   7. In covalent bonding, two or more atoms share electrons so that each atom becomes more stable; atoms are most stable in one of three states
      a. The outer electron shell is completely filled
      b. The outer electron shell contains eight valence electrons, according to the octet rule
      c. The outer electron shell is completely empty
   8. In covalent bonding, atoms may share any number of electrons so that each atom in the molecule attains one of the stable states given above
      a. For example, the hydrogen molecule (H_2) consists of two hydrogen atoms, each bringing one electron to the covalent bond; the outer electron shell of each atom becomes completely filled with two electrons
      b. The oxygen molecule (O_2) consists of two oxygen atoms, each bringing two electrons to the covalent bond; the two atoms share four electrons and the outer shell of each atom becomes completely filled with eight electrons

B. Types of covalent bonds
1. Scientists characterize covalent bonds as single, double, or triple, according to the number of pairs of electrons shared between the atoms
   a. Two atoms held together by one shared pair of electrons form a single bond, shown by drawing a single line between the two atoms (for example, H — H)
   b. Atoms sharing two pairs of electrons form a double bond, shown by drawing a double line between the atoms (such as O = O)
   c. Atoms sharing three pairs of electrons from a triple bond, shown by drawing three lines between the atoms (for example, N ≡ N)
2. Scientists also characterize covalent bonds according to whether or not the electrons are shared equally among the atoms; in a given covalent molecule, one atom may attract the shared electrons more strongly than the other
   a. When the atoms do not share the electrons equally, the atom that attracts the electrons more strongly will assume a partial negative charge; the atom that attracts the electrons less strongly will assume a partial positive charge
   b. Such a molecule (one with a partial electric charge on each end) is called a *polar molecule;* the atoms share a *polar covalent bond*
   c. If two atoms share electrons equally, no partial electric charge accumulates around either atom
   d. Atoms bound by equal sharing of electrons are called nonpolar molecules; the atoms share a nonpolar bond
3. Some chemical bonds are neither purely covalent nor purely ionic and fall somewhere between purely covalent and purely ionic
   a. These bonds involve only a partial transfer of an electron from one atom to the other
   b. The degree of unequal sharing depends on the electronegativity of the atoms involved

C. Properties of covalent compounds
1. Covalent compounds usually are gases, liquids, or solids with low melting points
2. Molecules in covalent compounds have no electrical charge; they are electrically neutral
3. A covalent compound does not dissolve in water; it retains its integral identity
4. Because covalent compounds do not dissolve to produce ions in water, they are poor conductors of electricity

IV. Metallic Bonds

A. General information
1. A **metallic bond**—a type of bond unique to elemental metals—is formed by the attraction of free negatively charged electrons to positively charged atomic centers
2. Scientists picture metal as a closely packed group of cations surrounded by a sea of constantly moving electrons

        a. The bonding electrons are not tied to any nucleus but move freely from one atom to another
        b. The strong electrical attractions between the positive cations and the negatively charged electrons hold the metal together
    3. Metallic bonding occurs mostly in pure elemental metals, such as silver, copper, sodium, iron, or potassium
B. Properties of metallic bond compounds
    1. These compounds have high melting points and do not dissolve in water
    2. Metallic bond compounds have an equal number of positive and negative charges and therefore are electrically neutral
    3. Because of their large number of wandering electrons, metallic bond compounds conduct electricity easily

V. Compounds and Chemical Formulas

A. General information
    1. A *compound* is a mixture of atoms from two or more elements
    2. The properties of a compound depend on numerous factors, including the elements present, the number of atoms of each element in one molecule of the compound, and the order in which the atoms combine to form the molecule
    3. The same atoms may combine in different ways to produce different shapes and sizes of molecules
        a. For example, hydrogen and oxygen may combine to form water (H_2O) — one oxygen atom and two hydrogen atoms
        b. Hydrogen and oxygen may also combine to form peroxide (H_2O_2) — two oxygen atoms and two hydrogen atoms
    4. A **chemical formula** — a symbolic representation of the atoms or elements that combine to form a compound — includes a combination of letters and numbers unique to each compound
        a. The letters in a formula indicate the elements in the compound
        b. The numbers (written as small subscripts to the right of the letters) indicate how many atoms of a particular element are contained in one molecule of the compound; the absence of a number after a letter indicates that only one atom of that element is present in the compound
            (1) The chemical formula NaCl indicates that a molecule of sodium chloride contains one atom of sodium and one atom of chlorine
            (2) The formula H_2SO_4 indicates that a molecule of sulfuric acid contains two atoms of hydrogen, one atom of sulfur, and four atoms of oxygen
    5. A compound composed of only two different elements is called a binary compound

B. Rules for naming compounds
    1. Compounds usually are named according to the order in which the elements appear in the periodic table groups, starting with the lowest numbered group and ending with the highest numbered group

Common Polyatomic Ions

NAME OF ION	FORMULA WITH ELECTRICAL CHARGE	EXAMPLE OF COMPOUND
Ammonium	NH_4^{+1}	Ammonium chloride (NH_4Cl)
Bicarbonate	HCO_3^{-1}	Sodium bicarbonate ($NaHCO_3$)
Hydroxide	OH^{-1}	Sodium hydroxide (NaOH)
Nitrate	NO_3^{-1}	Nitric acid (HNO_3)
Phosphate	PO_4^{-3}	Sodium phosphate (Na_3PO_4)
Sulfate	SO_4^{-2}	Sulfuric acid (H_2SO_4)

2. Binary compounds are named by stating the element from the lowest numbered periodic table group (usually a metal) first, then stating the remaining element (usually a nonmetal) second but changing the final syllable to the suffix -ide
   a. Silver chloride (AgCl) is written with silver (group IB) first and chlorine (group VIIA) second
   b. Copper sulfide (CuS) is written with copper (group IB) first and sulfur (group VIA) second
3. Scientists name ionic compounds by combining the names of the elements in the compound, referring to the positively charged ion first
   a. The usual procedure is to write the name of the positively charged ion (using its atomic name), then adding the name of the negatively charged ion (using its atomic name but substituting the ending with the suffix -ide)
   b. For example, the compound formed when sodium (Na^+) and chlorine (Cl^-) combine is called sodium chloride (NaCl)
4. Compounds also include prefixes (mono-, di-, and tri-) based on the number of atoms in each molecule
   a. Carbon monoxide (CO) contains one oxygen atom
   b. Carbon dioxide (CO_2) contains two oxygen atoms
   c. Phosphorous tribromide (PBr_3) contains three bromine atoms
5. Sometimes, a group of atoms assumes an overall electrical charge and behaves as a single ion; these charged particles (called **polyatomic ions**), have names fixed by convention
   a. A polyatomic ion (also called a *radical*) is a group of atoms that are held together by strong covalent bonds and that bears a net positive or negative charge just as if it were a single ion (see *Common Polyatomic Ions* for examples)
   b. A polyatomic ion can go through chemical reactions yet not come apart
6. Some molecules contain two or more polyatomic ions; when writing these compounds, scientists place parentheses around the polyatomic ion and follow the closing parenthesis with the subscript number that tells how many polyatomic ions are in the molecule

Compounds and Chemical Formulas 33

Comparing Formulas

Chemical formulas can be written in various ways, depending on the specific information—elements, number of atoms, or type of bonds—the scientist wishes to convey. The chart below compares the standard formulas for various compounds using the structural, condensed structural, molecular, and empirical methods. Within structural formulas, — indicates a single covalent bond, = indicates a double covalent bond, and ≡ indicates a triple covalent bond.

NAME OF COMPOUND	STRUCTURAL FORMULA	CONDENSED STRUCTURAL FORMULA	MOLECULAR FORMULA	EMPIRICAL FORMULA
Water	H — O — H	H_2O	H_2O	H_2O
Carbonic acid	$\begin{array}{c} O \\ \parallel \\ HO - C - OH \end{array}$	$HOCO_2H$	H_2CO_3	H_2CO_3
Ethane	$\begin{array}{c} H \quad H \\ \mid \quad \mid \\ H - C - C - H \\ \mid \quad \mid \\ H \quad H \end{array}$	CH_3CH_3	C_2H_6	CH_3
Acetylene	H — C ≡ C — H	CHCH	C_2H_2	CH

    a. Magnesium hydroxide is written $Mg(OH)_2$, indicating that two hydroxide ions are in each molecule
    b. Ammonium carbonate is written $(NH_2)_2CO_3$, which indicates two ammonium ions are in each molecule

C. **Types of chemical formulas**
    1. A *structural formula* shows how all the atoms in the molecule are attached to each other; this type of formula commonly is used in organic chemistry and biochemistry
    2. Organic chemists also use the *condensed structural formula* when atoms appear more than once in the molecule; subscripts help shorten the formula, but the structure still shows how the atoms are attached to each other
    3. Inorganic chemists generally use the *molecular formula,* which shows the actual number of atoms in a molecule, but not the molecule's structural arrangement
    4. An *empirical formula* shows the relative numbers of atoms in a molecule, but not necessarily the exact number present
    5. The molecular formula and the empirical formula for a given compound may be the same or different (see *Comparing Formulas* for examples)
    6. Structural formulas demonstrate different compounds that may have the same molecular and empirical formulas
        a. This is the case with complex organic molecules with the same molecular and empirical formulas but with different structures; for example, butane has a molecular formula of C_4H_{10} but an empirical formula of C_2H_5

b. Structurally, butane can be expressed according to the two formulas shown below

n-butane

```
    H   H   H   H
    |   |   |   |
H—  C — C — C — C  —H
    |   |   |   |
    H   H   H   H
```

or

iso-butane

```
    H   H   H
    |   |   |
H—  C — C — C  —H
    |   |   |
    H   |   H
        H — C — H
            |
            H
```

D. Formula weight
1. The **formula weight** of a compound is the sum of the atomic weights of all the atoms in its chemical formula
2. The formula weight for water (H_2O) is 18 atomic mass units (amu)
    a. The atomic weight of the two hydrogen atoms is two times the atomic weight of one hydrogen atom; 2×1 amu $= 2$ amu
    b. The atomic weight of the one oxygen atom is 16 amu
    c. 2 amu + 16 amu = 18 amu

Study Activities

1. Explain how ionic, covalent, and metallic bonds are formed.
2. Describe the properties of ionic, covalent, and metallic compounds.
3. Define a binary compound and give three examples.
4. Calculate the formula weights of carbon monoxide and carbon dioxide.

6

Chemical Reactions and Equilibrium

Objectives

After studying this chapter, the reader should be able to:
- State the law of conservation of mass.
- Describe the steps for balancing chemical equations.
- List the rules for assigning oxidation numbers to elements.
- Describe the steps for balancing an oxidation-reduction equation.
- List four factors that will affect the rate of a chemical reaction.
- Define catalyst, and explain the effect of a catalyst on the rate of a chemical reaction.
- Define chemical equilibrium.

I. Chemical Reactions

A. General information
1. A *chemical reaction* involves the interaction of atoms or molecules to form new compounds
2. For a chemical reaction to occur, different atoms or molecules must move close enough for the outer-shell electrons to interact; the energy generated by atoms or molecules moving closer together is called **kinetic energy**
3. Also, the atoms or molecules must collide with sufficient kinetic energy to overcome the repelling forces of the electrons surrounding the nucleus of the atoms involved; the total amount of kinetic energy needed to achieve successful collision and reaction is called the **activation energy**
4. Scientists describe the outcome of a chemical reaction using a shorthand method called a *chemical equation*

B. Reaction rate
1. The term **reaction rate** describes how quickly a chemical reaction occurs under given circumstances, such as at a certain temperature or under particular light conditions
2. Various factors can affect the reaction rate
   a. The identity of the reacting substances **(reactants)** influences the reaction rate; some reactants react more or less quickly than others (for example, nitric oxide reacts very quickly with atmospheric oxygen to form nitrogen dioxide, whereas carbon monoxide reacts very slowly with oxygen to form carbon dioxide)

b. The concentration of the reactants influences the reaction rate (for instance, the greater the number of reactant atoms, the more likely the atoms will collide and produce a reaction)
c. The surface area of a solid reactant influences the reaction rate (for instance, the greater the surface area available for interaction, the higher the reaction rate)
d. Temperature influences the reaction rate (for instance, atoms and molecules tend to gain more kinetic energy with increased temperatures, thereby increasing the probability that there will be sufficient energy to provide the necessary activation energy for the reaction to occur)
3. Because many reactions occur extremely slowly, scientists sometimes add substances called **catalysts** to help speed the process
   a. A catalyst reduces the activation energy required to initiate and sustain a reaction
   b. The catalyst itself is never changed by the reaction

C. Exothermic reactions
1. Any group of atoms or molecules possesses stored energy *(potential energy)* that is associated with the attractions and repulsions among the various particles
2. An **exothermic reaction** is a chemical reaction that releases energy; in many cases, the released energy appears as heat, which can be measured
3. In this type of reaction, the products have less potential energy than the reactants do
4. Once initiated, an exothermic reaction is self-sustaining and continuous
5. Examples of exothermic reactions include the explosion of dynamite and the burning of a candle

D. Endothermic reactions
1. An **endothermic reaction** is a chemical change that requires a continuous input of energy for the reaction to begin and to continue
2. In this type of reaction, the potential energy of the products is greater than that of the reactants; in other words, energy is absorbed by the reaction
3. If energy input ceases, the reaction will stop
4. Examples of endothermic reactions include photosynthesis in plants and electrolysis of water molecules; in photosynthesis, the continuous supply of energy from the sun enables the conversion of carbon dioxide and water to glucose (sugar) and oxygen

E. Oxidation-reduction reactions
1. Many chemical reactions involve the complete or partial transfer of electrons from one atom to another
2. Scientists use the term **oxidation-reduction reaction** (or *redox*) to describe any reaction involving the transfer of electrons between reactants
   a. **Oxidation** is the loss of electrons by an atom
   b. **Reduction** is the gain of electrons by an atom
3. In any oxidation-reduction reaction, the number of electrons gained by one reactant must always equal the number of electrons lost by another reactant
   a. An **oxidizing agent** is any reactant that gains electrons

Sample Oxidation Numbers

Oxidation numbers provide a means of tracking the transfer of electrons in chemical reactions. The specific numbers are assigned according to a set of rules. This chart lists elements commonly involved in oxidation-reduction reactions along with their assigned numbers and examples of the compounds they help form.

ELEMENT	OXIDATION NUMBER	EXAMPLE OF COMPOUND
GROUP IA		
H	+1	Hydrogen chloride (HCl)
Na	+1	Sodium chloride (NaCl)
K	+1	Potassium chloride (KCl)
GROUP IIA		
Mg	+2	Magnesium chloride ($MgCl_2$)
Ca	+2	Calcium chloride ($CaCl_2$)
TRANSITION METALS		
Fe	+2, +3	Iron(II) chloride ($FeCl_2$), Iron(III) chloride ($FeCl_3$)
GROUP IIIA		
Al	+3	Aluminum chloride ($AlCl_3$)
GROUP VIA		
O	−2	Ferrous oxide (FeO_2)
S	−2	Sodium sulfide (Na_2S)
GROUP VIIA		
Cl	−1	Sodium chloride (NaCl)
Br	−1	Sodium bromide (NaBr)

     b. A ***reducing agent*** is any reactant that loses electrons
     c. In the equation $Ca + Cl_2 \rightarrow Ca^{2+} + 2Cl^-$, the calcium (Ca) loses two electrons and the chlorine (Cl) gains two electrons; Ca, the element oxidized, is the reducing agent and Cl, the element reduced, is the oxidizing agent
4. Scientists use **oxidation numbers** (also called oxidation *states, levels,* or *valences*) to track the transfer of electrons in chemical reactions; the oxidation number of an atom is equal to the number of positive or negative charges that the atom carries (see *Sample Oxidation Numbers*)
5. Oxidation numbers are assigned according to specific rules
     a. Any atom in its free state (without a charge) has an oxidation number of zero
     b. Any monoatomic ion (an ion consisting of only one atom) has the same oxidation number as its ionic charge
     c. In a polyatomic ion (an ion consisting of more than one atom), the algebraic sum of the oxidation numbers of all the atoms is the same as the overall charge on the ion
     d. In a covalent bond, oxidation numbers are assigned to each atom as if electrons were transferred to the more electronegative atom

e. In an electrically neutral molecule, the algebraic sum of the oxidation numbers of all the atoms is zero
   f. Oxygen typically has an oxidation number of -2
   g. Hydrogen typically has an oxidation number of $+1$
   h. Certain metals, especially transition elements, can have more than one ionic form; hence, each ionic form will have a different oxidation number (for example, iron can form two different positive ions; the ferric ion has an oxidation number of $+3$, and the ferrous ion has an oxidation number of $+2$)
   i. The oxidation number of most other elements varies with the compound
 6. Oxidation numbers also may be assigned according to the periodic group to which an element belongs
   a. Any element in Group I has an oxidation number of $+1$
   b. Any element in Group II has an oxidation number of $+2$
   c. Any element in Group VII has an oxidation number of -1
   d. Any element in Group VI has an oxidation number of -2
   e. Any element in Group V has an oxidation number of -3

II. Chemical Equations

A. General information
 1. Scientists can predict and measure the results of a chemical reaction by knowing the unique characteristics of the atoms or molecules involved
 2. Scientists write the results of chemical reactions using a shorthand method called a *chemical equation;* a chemical equation must describe the chemical reaction as it actually occurs
 3. All chemical reactions obey the **law of conservation of mass,** which states that matter is neither created nor destroyed in any ordinary chemical reaction; the sum of the masses of the reactants always equals the sum of the masses of the substances *(products)* formed
 4. Because chemical reactions follow the law of conservation of mass, all chemical equations should balance, accounting for the total number of atoms on both sides of the equation
 5. In a chemical equation, the formula for the reactants appears on the lefthand side of the equation, and the formula for the products appears on the righthand side of the equation
 6. An arrow pointing toward the products indicates the words "yields" or "produces"; for example, $A + B \rightarrow C + D$ (see *Common Symbols for Writing Equations*)

B. Seven-step method for balancing chemical equations
 1. Write the correct chemical formula for each reactant and for each product; because the known formula for a compound does not change, *never* alter the correct chemical formulas for the reactants or the products
 2. Choose the most complex compound (the one with the greatest number of atoms) from either the products or the reactants; give this compound a temporary coefficient of 1

Common Symbols for Writing Equations

This chart includes some of the most commonly used symbols for writing chemical equations along with an explanation of their use.

SYMBOL	MEANING
+	A plus sign is used between two formulas to indicate the reactants combined or the products formed.
→	An arrow is used to separate reactants (on the left) from products formed (on the right). The arrow indicates the direction of the change. It can be read as "yields" or "produces."
⇌	A double arrow may be used in place of an arrow to show that the reaction can occur in both directions. It also may be used to indicate that the reaction is at equilibrium—that is, the forward and reverse reactions are occurring at the same rate.
(s)	A lower-case "s"—placed within parentheses after the element symbol or formula—may be used to indicate that a substance is a solid.
(l)	A lower-case "l"—placed within parentheses after the element symbol or formula—indicates that the substance is a liquid.
(g)	A lower-case "g"—placed within parentheses after the element symbol or formula—indicates that the substance is a gas.
(aq)	The lower-case letters "aq"—placed after the element symbol or formula—indicates that the substance is aqueous, or dissolved in water.

    a. A *coefficient* is an arbitrary number placed in front of each molecule of the equation to satisfy the law of conservation of mass
    b. Molecules with no coefficient are assumed to have a coefficient of 1
    c. For example, the equation $H_2SO_4 + BaCl_2 \rightarrow BaSO_4 + 2HCl$ indicates that one molecule of H_2SO_4 plus one molecule of $BaCl_2$ combine to yield one molecule of $BaSO_4$ and two molecules of HCl
3. Select any one element from the most complex compound; assign coefficients to molecules on either side of the equation, then balance the number of atoms for this single element so that the number of atoms for the element is the same on both sides of the equation
4. Next, select another element from the most complex compound and balance it according to the same manner as in step 3; change the coefficient of the other molecules as needed to balance each of the elements
    a. Continue selecting elements and assigning or changing coefficients as necessary so that the total number of atoms for each element is the same on both sides of the equation
    b. Never change the subscripts inside the formulas themselves; only change the coefficients as many times as necessary to balance the equation
5. As a general rule, balance the oxygen atoms last
6. If a polyatomic ion is included in the molecule, consider it a single unit, if it remains unchanged in the chemical reaction
7. After balancing the equation, add up the total number of atoms for the different elements on each side of the equation; the number of atoms of each different element should be the same on both sides of the equation

C. Five-step method of balancing redox equations
1. Write the correct formulas for both the reactants and the products
2. Write the oxidation number for each element, remembering that the oxidation number of uncombined elements is zero
3. Determine which elements changed their oxidation number; show the electrons gained or lost for each element
4. Remember that the number of electrons lost by one reactant must be the same as the number of electrons gained by another reactant; assign coefficients to reactants and products to balance the numbers of electrons lost with the number of electrons gained
5. Balance the remaining parts of the equation by assigning coefficients
6. Check the final, balanced equation to ensure that the total number of atoms of each different element is the same on both sides of the equation

III. Chemical Equilibrium

A. General information
1. In most chemical reactions, the reaction stops before all of the reactants are used up; at some point the products begin to react with each other to reform the reactants—that is, the reaction begins to reverse itself
2. **Chemical equilibrium** is the dynamic state in which both the forward and reverse reactions occur simultaneously at exactly equal rates
3. For example, when sugar is first added to iced tea, it will dissolve until the tea is unable to dissolve any more; at that point, the sugar will begin to crystallize or come out of solution
4. As in the human body, a chemical equilibrium between carbon dioxide and carbonic acid exists in the blood; this equilibrium allows the body to rid itself of the waste product CO_2 via the lungs

B. Factors that affect chemical equilibrium
1. Changing the concentration of a reactant or product will disturb the equilibrium; for example, adding more reactants may force the reaction in a forward direction temporarily
2. Changing the temperature also will alter the equilibrium
   a. Increasing the temperature causes endothermic reactions to move in a forward direction
   b. Reducing the temperature causes exothermic reactions to move in a forward direction
3. Catalysts do not affect chemical equilibrium; rather, they alter the rate of both the forward and the reverse reactions equally

Study Activities

1. Describe how a chemical reaction occurs, and explain the significance of activation energy.
2. Compare and contrast exothermic and endothermic reactions, and give several examples of each.
3. Explain what happens in an oxidation-reduction reaction using tarnished silver or rusted iron as examples.
4. Describe the function of a chemical equation, and explain why it must be balanced.
5. Identify two factors that affect chemical equilibrium, and list their possible effects.

7

Solutions

Objectives

After studying this chapter, the student should be able to:
- Define mole.
- Define molarity, molality, and normality.
- Define solute and solvent.
- State the three different ways to express percent solution.

I. The Mole

A. General information
1. Scientists measure a substance in two ways: by weighing the substance on a scale to learn its mass (weight) and by counting the number of particles in the substance
    a. Mass is commonly measured on a metric scale that shows weight in grams
    b. The number of particles in a given substance is indicated by a unique counting unit called a **mole**
2. A *mole* is a standard number that allows scientists to express the number of particles in one substance in relation to the number of particles in other substances; it is equivalent to the amount of substance that contains as many particles (atoms, molecules, electrons, or ions) as the number of atoms in exactly 12 grams of the carbon-12 isotope (^{12}C)
3. The term *mole* is similar in concept to the terms *dozen* (which always means 12), *gross* (which always means 144), and *score* (which always means 20)
4. One mole of a substance is equal to 6.02×10^{23} particles
5. The number 6.02×10^{23} is called **Avogadro's number**, named after the nineteenth-century Italian physicist whose experiments with gases led to its discovery
6. The mole is a unique measuring unit because it allows scientists to measure mass and number at the same time
    a. One mole of atoms of any element always has a weight in grams equal to the atomic weight of that element; the **molar weight** of an atom is the weight in grams that is equal to the atomic weight in atomic mass units

b. One mole of molecules of any compound always has a weight in grams equal to the formula weight of that compound; the molar weight of a compound is the weight in grams that is equal to the formula weight in atomic mass units

B. The mole in chemical equations
1. A balanced chemical equation is a representation of the reactions that take place on the molecular level as well as on the molar level
2. Because the weight of a mole of any compound is the same as its formula weight, the mole concept may be used to calculate how many grams of each reactant are needed to produce any given weight of product

II. Mixing Solutions

A. General information
1. A *solution* is a homogeneous mixture of two or more substances dispersed as molecules, atoms, or ions and with uniform properties; it may be a gas, a liquid, or a solid
2. The constituent parts of a solution include a solvent and one or more solutes
    a. The component present in the greatest amount is called the **solvent**
    b. The component (or components) present in the least amount is called the **solute**
3. In a solution, one or more solutes is dissolved in a solvent
4. To ensure that chemical reactions occur in solutions, scientists must place a precise number of moles of each reactant in a given volume of solute
5. Scientists can prepare solutions in various concentrations
    a. The *concentration* refers to the amount of solute in the solution
    b. Depending on the amount of solute, a solution can range from highly concentrated to extremely dilute
        (1) A *concentrated solution* is one with a large amount of solute in solution
        (2) A *dilute solution* is one with a relatively small amount of solute in solution
6. Because of the need to quantitate how concentrated or dilute a solution is, scientists express concentration in precise units based on various formulas; these formulas provide the exact concentration of solute in the solution but express the concentration using different terms (percent solution, molarity, molality, and normality), depending on the specific unit of measure

B. Percent solution
1. A percent solution is a mixture of solute in a given solvent expressed in terms of the precise weight or volume—or combination of weight and volume—of its constituent parts
2. A **percent** *(volume/volume)* **solution** is used to express the volume of solvent in a given volume of solute; it is a convenient way of expressing a liquid-liquid solution
    a. The percent (volume/volume) is equivalent to the milliliters of solute in a total volume of 100 milliliters of solution multiplied by 100%

b. Computation involving this type of solution are expressed as follows:

$$\% \text{ volume/volume} = \frac{\text{ml of solute}}{100 \text{ ml of solution}} \times 100\%$$

3. A percent *(weight/weight)* solution is used to express the weight of solute in a given weight of solution; it is another way of expressing a liquid-liquid solution except that it substitutes grams for milliliters
   a. The percent (weight/weight) solution is equivalent to the grams of solute in a total of 100 grams of solution multiplied by 100%
   b. Computations involving this type of solution are expressed as follows:

$$\% \text{ weight/weight} = \frac{\text{g of solute}}{100 \text{ g of solution}} \times 100\%$$

4. A percent *(weight/volume)* solution is commonly used in laboratory settings because of the ease of preparation
   a. The percent (weight/volume) is equivalent to the grams of solute in a total of 100 milliliters of solution multiplied by 100%
   b. Computations involving this type of solution are expressed as follows:

$$\% \text{ weight/volume} = \frac{\text{g of solute}}{100 \text{ ml of solution}} \times 100\%$$

C. Molarity

1. The **molarity** of a solution is the number of moles of solute in 1 liter of solution; molarity = moles of solute/liters of solution
2. The common abbreviation for the molarity of a solution is M; for example, a 3-molar solution of sodium chloride is labeled 3M NaCl
3. The molarity of a solution varies with the temperature of the solution because the volume of any given solution usually expands with high temperatures and contracts with low temperatures

D. Molality

1. The **molality** of a solution is the number of moles of solute dissolved in 1 kilogram of solvent; molality = moles of solute/kilograms of solvent
2. The molality unit is based on a fixed weight of solvent, whereas the molarity unit is based on a fixed volume of total solution
3. A molal solution typically uses a solvent other than water, such as alcohol, acetone, or chloroform
4. The molality of a solution remains the same regardless of the temperature of the solution because mass does not depend on temperature; scientists use molality in experiments requiring a constant concentration of a substance despite changes in temperature

E. Normality

1. Normality is a concentration unit used in the reaction between acids and bases; it also sometimes is used in reactions between ionic compounds and in oxidation-reduction reactions
2. The **normality** of a solution is the number of equivalents of solute in 1 liter of solution; normality = equivalents of solute/liters of solution
3. An **equivalent** is a specific quantity of reactant needed to supply 1 mole of reacting units in a chemical reaction

a. For acids, an equivalent is the amount of acid that will supply 1 mole of H^+ ions on complete dissociation; for bases, an equivalent is the amount of base that will supply 1 mole of OH^- ions on complete dissociation; a **gram equivalent** of acid is the weight of acid in grams that will supply 1 mole of H^+ ions; a gram equivalent of base is the weight of base in grams that will supply 1 mole of OH^{-1} ions
   (1) For example, the formula for sulfuric acid is H_2SO_4; each molecule of acid completely dissociates to yield two molecules of H^+ and one molecule of SO_4^-
   (2) One mole of sulfuric acid contains two equivalents of sulfuric acid because one mole of sulfuric acid contains two moles of H^+; 1 mole H_2SO_4 = 2 equivalents H_2SO_4
   (3) Therefore, one equivalent of sulfuric acid is equal to ½ mole of sulfuric acid; ½ mole H_2SO_4 = 1 equivalent H_2SO_4
b. For ionic compounds, an equivalent is the amount of substance that will supply one mole of negative charge or one mole of positive charge upon complete dissociation; a gram equivalent of an ionic compound is the weight of ionic compound in grams that will supply one mole of negative or positive charge
   (1) For example, the formula for calcium nitrate is $Ca(NO_3)_2$; each molecule of calcium nitrate dissociates to yield one molecule of Ca^{+2} and two molecules of $(NO_3)^{-1}$
   (2) One mole of calcium nitrate contains two moles of positive charge and two moles of negative charge; 1 mole $Ca(NO_3)_2$ = 2 equivalents of $Ca(NO_3)_2$
   (3) Therefore, one equivalent of calcium nitrate is equal to ½ mole of calcium nitrate; ½ mole $Ca(NO_3)_2$ = 1 equivalent $Ca(NO_3)_2$
c. For both oxidizing and reducing agents, an equivalent is the amount of substance that will accept one mole of electrons in an oxidation-reduction reaction

F. Dilution
1. Although solutions are more easily, accurately, and precisely prepared at higher concentrations, scientists generally work in smaller concentrations in the laboratory
2. To prepare solutions for laboratory work, scientists must dilute high-concentration solutions to low-concentration solutions
3. When diluting solutions from a given concentration to a lower concentration, scientists follow this simple formula:
$$C_1V_1 = C_2V_2$$
   a. C_1 and V_1 are the initial concentration and volume
   b. C_2 and V_2 are the final concentration and volume
4. This simple formula may be used as long as all of the concentration and volume units involved in the calculation are the same
5. The concentration of an extremely dilute solution may be expressed in several ways, such as parts per billion (ppb) or milligrams per deciliter (mg/dl) (for a more complete list of commonly used units, see *Common Concentration Units for Dilute Solutions,* page 46)

Common Concentration Units for Dilute Solutions

Scientists express the concentration of dilute solutions using the following units.

CONCENTRATION UNIT	MEANING OF UNIT		ABBREVIATED SYMBOL
Percent solution (volume/volume)	$\dfrac{\text{milliliters of solute}}{\text{100 milliliters of solution}}$	or ml/100 ml	% (v/v)
Percent solution (weight/weight)	$\dfrac{\text{grams of solute}}{\text{100 grams of solution}}$	or g/100g	% (w/w)
Percent solution (weight/volume)	$\dfrac{\text{grams of solute}}{\text{100 ml of solution}}$	or g/100 ml	% (w/v)
Milligrams percent or milligrams per deciliter	$\dfrac{\text{milligrams of solute}}{\text{100 milliliters of solution}}$	or mg/100 ml	mg% *or* mg/dl
Parts per million	$\dfrac{\text{milligrams of solute}}{\text{liters of solution}}$	or mg/L	ppm
Parts per billion	$\dfrac{\text{milligrams of solute}}{\text{1,000 liters of solution}}$	or mg/1,000L	ppb
Parts per billion (gases)	$\dfrac{\text{microliters of solute}}{\text{liters of air}}$	or μl/L	ppb

Study Activities

1. Explain molar weight and the use of the mole concept in chemical reactions.
2. Describe the effect of temperature on molarity and molality.
3. Outline a step-by-step method for determining equivalents.
4. Calculate the molarity, molality, and percent solution of a solution in which 10 grams of sodium hydroxide are dissolved in 1 liter of distilled water.
5. Calculate the molality of a solution in which 10 grams of sodium hydroxide are dissolved in 500 grams of water.

8

Solids

Objectives

After studying this chapter, the reader should be able to:
- Identify the two types of solids and tell how they differ from each other.
- Describe how molecular solids differ from ionic solids.
- Describe the four types of crystalline solids, giving examples of each.
- Explain the five ways scientists classify and describe crystals.
- Name and describe four different types of bonds found in crystalline solids.

I. The Structure of Solids

A. General information
1. A *solid* is a form of matter—a compound or an element whose molecules or atoms are tightly compacted, allowing little freedom of movement
2. Because of the arrangement of its molecules or atoms, a solid has a definite shape and volume that does not necessarily conform to the shape of a container
3. Scientists classify solids as either crystalline or amorphous, depending on the arrangement of the atoms or molecules

B. Crystalline solids
1. A ***crystalline solid*** consists of atoms, ions, or molecules arranged in a regularly repeating, three-dimensional pattern; common examples include quartz, table salt, diamonds, and snowflakes
2. The atoms, ions, or molecules of a crystalline solid occupy specific positions, called lattice points
3. As a rule, most solids are crystalline

C. Amorphous solids
1. An ***amorphous solid*** consists of a fixed, randomly nonrepeating arrangement of atoms or molecules; common examples include glass, rubber, and most plastics
2. Any solid that is not crystalline is considered amorphous
3. Amorphous solids have no definite freezing or melting points; the transition from solid to liquid or from liquid to solid occurs over a temperature range
4. Most amorphous solids do not dissolve in water; typically, they are brittle and break or shatter readily into fragments with curved, shell-like surfaces

5. Most amorphous solids will not conduct electricity
6. Many amorphous solids are supercooled liquids; for example, most glass is a supercooled liquid with molecules frozen into the solid form

D. Other ways of classifying solids
1. Scientists also classify solids by the kinds of forces that hold the atoms or molecules together; these forces determine the structure and properties of each type of solid (for example, solids held together by intermolecular forces differ from solids held together by ionic attraction)
2. **Molecular solids** are held together by relatively weak intermolecular forces, such as van der Waals forces and hydrogen bonding; molecular solids generally have lower melting points than ionic solids do
   a. *Intermolecular forces* are any forces of attraction between molecules
   b. **Van der Waals forces** are intermolecular forces that result from the temporary dispersion of the electrons in a molecule toward one end of the molecule
      (1) One end of the molecule momentarily gains electrons and the other end loses electrons; the portion of the molecule with more electrons assumes a slight negative charge; the portion of the molecule with fewer electrons assumes a slight positive charge
      (2) The negatively charged portion of a molecule attracts the positively charged portion of other molecules
      (3) The strength of the van der Waals forces between molecules increases as molecular size increases
   c. **Hydrogen bonding** is a special type of intermolecular bonding that occurs when the hydrogen in a molecule forms additional loose bonds with electronegative atoms in neighboring molecules
      (1) When hydrogen forms a molecule by bonding with another, more electronegative atom, the more electronegative atom tends to pull the hydrogen electron strongly toward itself
      (2) As a result of this unequal sharing of electrons within the molecule, the hydrogen atom assumes a partial positive charge
      (3) The positively charged hydrogen end of the molecule attracts the negatively charged end of other molecules, forming a temporary bond called a *hydrogen bond*
3. **Ionic solids** are held together by the strong forces of attraction between oppositely charged ions; ionic solids typically have higher melting points than molecular solids do

II. Crystals

A. General information
1. Crystals are solids composed of atoms, ions, or molecules arranged to occupy specific positions
   a. The center of each of these positions is called a *lattice point*
   b. The lattice points of a crystal are arranged in a repeating, geometric pattern to form the crystal structure

Physical Properties of Crystals

The chart below compares some of the physical properties of the four types of crystals—ionic, covalent, molecular, and metallic.

PHYSICAL PROPERTY	IONIC	COVALENT	MOLECULAR	METALLIC
Examples	Sodium chloride	Carbon, diamond	Carbon dioxide, water, iodine	Gold, iron
Unit at lattice point	$^+$ and $^-$ ions	Atoms	Atoms or molecules	Atoms or ions (or both)
Bonding forces	Electrostatic charges	Covalent bonds	Hydrogen or other type bonds	Electrostatic charges
Hardness	Hard	Hard	Soft	Soft to hard
Melting point	High	High	Low	Low to high
Conduction of heat and electricity	Extremely poor	Extremely poor	Extremely poor	Good

    2. The bonds between the atoms or molecules of a crystal may be ionic, covalent, hydrogen, or a combination of these and other forces
    3. In addition to describing crystals by their bond type, scientists characterize crystals according to their hardness, melting point, heat conduction, and conduction of electricity (see *Physical Properties of Crystals*)

B. Crystal structure

1. The basic repeating unit of crystals is a collection of geometrically arranged atoms, ions, or molecules called the *unit cell*
2. The size and shape of the unit cell depends on the size and shape of the atoms, ions, or molecules and how they readily pack together
3. Unit cells occur in only seven shapes—simple cubic, tetragonal, orthorhombic, rhomobohedral, monoclinic, triclinic, and hexagonal; every crystalline solid can be described in terms of one of these seven types (see *Types of Unit Cells,* page 50)
4. The repetition of unit cells forms the lattice structure that is characteristic of a crystalline solid
5. This knowledge of crystal structure results from **X-ray diffraction** studies
   a. In X-ray diffraction, scientists bombard a sample of matter with X-rays
   b. By studying how a crystalline solid scatters and changes the intensity of the X-rays, scientists can determine how the molecules or atoms are arranged in the crystal
6. Some elements exhibit crystal allotropy—that is, they assume different crystal forms under different conditions (some common examples include carbon, which forms graphite, coal, and diamonds; calcium, which is found in limestone, chalk and marble; and sulfur, which changes its physical state when heated, changing from a solid to liquid forms with varying viscosities)

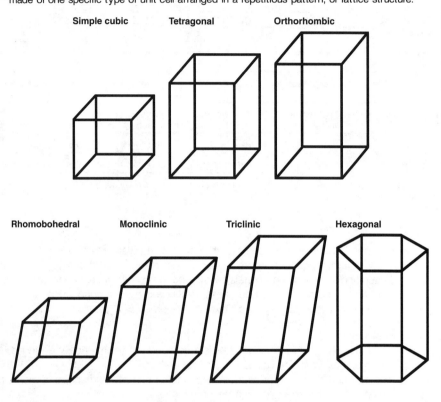

Types of Unit Cells

Scientists have identified only the seven types of unit cells illustrated below. Each crystal is made of one specific type of unit cell arranged in a repetitious pattern, or lattice structure.

Simple cubic Tetragonal Orthorhombic

Rhomobohedral Monoclinic Triclinic Hexagonal

C. **Types of crystals**
1. Scientists divide crystals into four different classes depending on the type of bond that holds the atoms or molecules together; these are ionic, covalent, molecular, or metallic
2. The bond type determines the characteristic properties of the crystal

D. **Ionic crystals**
1. *Ionic crystals* are composed of cations and anions arranged in unit cells so that each ion is surrounded by neighboring ions having the opposite charge
2. Usually, the anions and cations differ markedly in size
3. The forces of electrostatic attraction between the positive and negative ions holds the unit cells together
4. Ionic crystals are hard and brittle and have high melting and boiling points; they are poor conductors of heat and electricity
5. Common ionic crystals include sodium chloride, potassium chloride, and potassium nitrate

E. Covalent crystals
1. *Covalent crystals* are composed of atoms held together in an extensive three-dimensional network by extremely strong covalent bonds
2. Sometimes a single type of atom forms a covalent crystal; for example, the carbon atom forms two different covalent crystals—graphite and diamond
3. Quartz (SiO_2) is an example of two different atoms—silicon and oxygen—combined to form a covalent crystal
4. Covalent crystals are hard—in many cases, extremely hard; they have high melting points and conduct electricity and heat poorly

F. Molecular crystals
1. *Molecular crystals* are composed of molecules packed as closely together as their size and shape allow
2. The forces that bind these molecules—van der Waals forces or hydrogen bonding (or both)—generally are weaker than the bonds that hold ionic and covalent crystals
3. Molecular crystals are soft compared with other types of crystals; they have relatively low melting points and conduct electricity and heat poorly
4. Common molecular crystals include ice (solid H_2O), solid sulfur dioxide (SO_2), solid iodine (I_2), solid phosphorus (P_4) and solid sulfur (S_8)

G. Metallic crystals
1. *Metallic crystals* are composed of many metal atoms, all consisting of the same element and closely packed together in a dense crystalline structure
2. In metallic crystals, each lattice point is occupied by a positive metal ion (the metal atom without its valence electrons); the valence electrons form a sea of electrons that circulates freely about the fixed metal ions
3. This sea of freely moving electrons in metallic crystals enables metals to conduct electricity easily
4. Metallic crystals usually are strong, but they may vary from hard to soft depending on the element; they have variable melting points and are good conductors of both heat and electricity
5. Common metallic crystals include gold, silver, and iron

Study Activities
1. Outline the various methods of classifying solids.
2. Describe the unit cell and its function in a crystalline solid.
3. Describe and draw the structure of table salt, and explain its bond type, hardness, melting point, heat conduction, and conduction of electricity.

9

Liquids

Objectives

After studying this chapter, the reader should be able to:
- Describe six characteristic properties of liquids.
- Describe and compare solutions, colloids, and suspensions, and identify the characteristic properties of each.
- Discuss the effects of temperature on the solubility of a solid solute in a liquid solvent.
- Explain why ice floats on water.
- Explain why water is called the universal solvent.
- Describe an electrolytic solution.

I. Properties of Liquids

A. General information
1. A *liquid* is a compound or element whose molecules or atoms are relatively close together but have much freedom of movement
2. Because of the mobility of its molecules or atoms, a liquid has a definite volume and an indefinite shape
3. Liquids usually occur as compounds and rarely as elements
    a. Only a few elements—bromine, gallium, cesium, rubidium, and mercury—are liquid at room temperature
    b. Water—a pure compound—is liquid only under certain conditions
4. Scientists characterize liquids by measuring six different physical properties: viscosity, surface tension, density, compressibility, boiling point, and freezing point
    a. The strength of the forces of attraction between the molecules or atoms of the liquid determines the values of these physical properties
    b. The addition of energy, particularly in the form of heat, can alter the forces of attraction, thereby changing the physical property of the liquid

B. Viscosity
1. **Viscosity** is a measure of the ease of liquid flow; the higher the viscosity, the less easily the liquid flows (for example, vegetable oil has a higher viscosity than water and a lower viscosity than honey; it flows less easily than water and more easily than honey)
2. Liquids with strong intermolecular forces of attraction have greater viscosities than those with weak intermolecular forces of attraction

3. When the temperature of a liquid rises, its viscosity decreases; conversely, when the temperature of a liquid drops, its viscosity increases

C. **Surface tension**
 1. *Surface tension* is a measure of the resistance necessary to keep the the molecules at the surface of a liquid from expanding the liquid's surface area; this phenomenon is responsible for the ability of a drop of liquid to keep its shape
 2. The molecules within a liquid are attracted in all directions to surrounding molecules by intermolecular forces, such as van der Waals forces and hydrogen bonding; however, molecules on the liquid surface are attracted laterally and downward, but not upward
 3. This lack of balance in the forces of attraction causes surface molecules to crowd together and form a "skin" on the surface of the liquid (for example, surface tension causes water to bead up on a freshly waxed car; this same property makes a soap bubble round)
 4. The attraction of surface molecules to molecules within the liquid is called *cohesive force;* the strong attraction of surface molecules to unlike molecules outside the liquid is called *adhesive force*
 5. As the temperature of a liquid increases, its surface tension decreases
 6. *Wetting agents* or **surfactants** are chemicals that artificially decrease the surface tension of a liquid
    a. The molecules of the wetting agent attract the liquid's surface molecules and thus overcome its normal cohesive forces
    b. The resultant decrease in surface tension permits the liquid to spread out (for example, a detergent is considered a wetting agent)

D. **Density**
 1. The *density* of any substance is equal to its mass divided by its volume; $D = m/V$
 2. Scientists express the densities of solids as grams per cubic centimeter (g/cm^3), liquids as grams per milliliter (g/ml), and gases as grams per liter (g/L)
 3. The density of a substance changes with its temperature; generally, as the temperature of a substance increases, its density decreases
 4. Liquids have a higher density than gases and a lower density than solids; generally, liquids are approximately 1,000 times more dense than gases and 5% to 10% less dense than solids
 5. Water is a notable exception; liquid water at 0° C is 1.09 times more dense than solid water (ice)

E. **Compressibility**
 1. The **compressibility** of a substance is a measure of how much its volume decreases when external pressure is applied
 2. Liquids are about 10 times more compressible than solids, but they are far less compressible than gases; gases are about 100,000 times more compressible than solids
 3. This difference in compressiblity suggests that liquids have more "open spaces" or "holes" between molecules than do solids

F. Boiling point
1. On the surface of every liquid, a few molecules will always break through the surface tension, escape the liquid, and rise into the surrounding air as gaseous molecules
2. This change of state—from a liquid to a gas—in the surface molecules is called *evaporation*
3. The **vapor pressure** of a liquid is a measure of its tendency to evaporate
4. Every liquid has a characteristic vapor pressure that changes as the internal temperature of the liquid changes; generally, as the temperature of a liquid increases, its vapor pressure also increases
5. The *boiling point* of a liquid is the temperature at which its vapor pressure equals the pressure of the surrounding air; at this temperature, molecules within the liquid form gaseous bubbles that rise to the surface and escape into the surrounding air
    a. The addition of heat energy raises the temperature of a liquid; it also causes a drop in the intermolecular forces of attraction between liquid molecules as well as a drop in the liquid's viscosity and surface tension
    b. As the temperature of a liquid rises, its molecules become increasingly jostled, moving more and more rapidly; this allows more molecules to escape (evaporate) from the liquid's surface
    c. Every liquid has a characteristic boiling point
    d. The boiling point of a liquid changes with atmospheric pressure; as atmospheric pressure decreases, the boiling point also decreases
6. The amount of heat energy required to change 1 gram of a substance from a liquid to a gas at its boiling point is called its **heat of vaporization**
7. When heat energy is lost from a gas, molecules may return to their liquid state in a process called *condensation,* which is the opposite of evaporation

G. Freezing point
1. The *freezing point* of a liquid is the temperature at which it changes from a liquid to a solid; the terms *freezing point* and *melting point* are synonymous
2. As a liquid loses heat energy, its temperature drops; molecules move more and more slowly, and intermolecular forces of attraction predominate
3. If enough heat energy is lost, the liquid assumes the characteristics of a solid; for example, as heat is removed from water, it will freeze into a solid (ice)
4. Every liquid has a characteristic freezing point
5. The amount of heat energy required to change 1 gram of a substance from a solid to a liquid at its freezing (melting) point is called its **heat of fusion**

II. Water

A. General information
1. A water molecule (H_2O) contains one atom of oxygen covalently bound to two atoms of hydrogen
2. Water possesses several unique properties as a result of hydrogen bonding and the polarity of the water molecule
3. **Polarity** is a measure of inequality in the sharing of bonding electrons

a. When one atom in a molecule pulls the bonding electrons more strongly than other atoms in the molecule, the molecule becomes negatively charged at one end and positively charged at the other end
   b. Oxygen pulls bonding electrons much more strongly than does hydrogen; in a water molecule, the oxygen atom assumes a partial negative charge and the hydrogen atoms assume a partial positive charge

B. **Unique properties of water**
  1. Water is called the universal solvent because it dissolves both ionic compounds and polar molecular compounds; water usually cannot dissolve nonpolar molecules
  2. Water has an unusually high melting point (0° C) and boiling point (100° C at 1 atm pressure) for a molecule of its size; most compounds of comparable molecular weight (water has a molecular weight of 18) are gases at room temperature, such as methane (molecular weight of 16 and a boiling point of −164° C) and ammonia (molecular weight of 17 and a boiling point of −33° C)
     a. Water's high boiling point indicates that a greater than normal amount of heat energy is required for gaseous molecules to escape from the liquid
     b. Because water molecules are attracted to each other by hydrogen bonding as well as by van der Waals forces, a greater amount of heat energy is necessary to separate water molecules from each other; consequently, water has a high heat of vaporization
     c. Water's high melting point indicates that a greater than normal amount of heat energy is required to convert from solid to liquid at the same temperature
     d. Because solid water (ice) is a crystalline lattice held by hydrogen bonds, a greater than normal amount of heat energy is required to melt water; consequently, water has a high heat of fusion
  3. Unlike most substances, water has a lower density when solid than when liquid; the unusually low density of ice causes it to float on liquid water
     a. Usually, the density of a substance increases as the temperature decreases; however, as water cools, its density increases only until it reaches a maximum density level (at approximately 4° C)
     b. Then, as the water cools further and freezes (at 0° C), its density actually *decreases*
  4. Water has a uniquely high surface tension because the surface molecules are bound to the liquid by hydrogen bonding as well as the usual van der Waals forces
  5. Compared with other liquids, water has a high specific heat
     a. The **specific heat** of a liquid is the amount of heat required to raise the temperature of 1 gram of liquid one degree Celsius
     b. The higher the specific heat of a substance, the less its temperature will change when it absorbs a given amount of heat
     c. As a result of its high specific heat, water can easily absorb heat or give off heat to surrounding substances; consequently, water helps keep the temperature of its surroundings relatively constant

Comparing Solutions, Colloids, and Suspensions

This chart reviews the properties of the three types of liquid mixtures—solutions, colloids, and suspensions.

PROPERTY	SOLUTIONS	COLLOIDS	SUSPENSIONS
Particle size	<1 nanometer (nm)	1 to 1,000 nm	>1,000 nm
Filtration	Passes through filters and membranes	Passes through filters but not membranes	Stopped by filters and membranes
Visibility of dissolved particles	Invisible	Visible with an electron microscope	Visible to eye or in a light microscope
Motion of dissolved particles	Molecular motion	Brownian movement	Downward movement by gravity
Effects of light	Transparent, no Tyndall effect	May be transparent, but often translucent or opaque; Tyndall effect observed	Often opaque, but may be translucent

III. Liquid Mixtures

A. General information

1. Liquid mixtures may be homogeneous or heterogeneous
    a. A *homogeneous mixture* is one having uniform properties throughout the mixture
    b. A *heterogeneous mixture* is one having non-uniform properties
2. Such mixtures may occur in one of three forms: liquid solution, colloid, or suspension (see *Comparing Solutions, Colloids, and Suspensions* for a quick review of the properties of these mixtures)
3. A *liquid solution* is a homogeneous mixture of two or more kinds of atoms, molecules, or ions
    a. Its constituents include a solvent (the liquid present in the greatest volume) and a solute (the substance dissolved in the solvent)
    b. The solute may be another liquid, a solid, or a gas; its particle size is quite small—generally smaller than 1 nanometer (nm) in diameter (1 nm = 10^{-9} meters)
4. A **colloid** is a heterogeneous mixture that sometimes, but not always, has a liquid solvent; in a colloid, the particle size of the solute is significantly greater than that in a solution (the size of solute particles in a colloid usually is between 1 nm and 1,000 nm)
5. A **suspension** is a heterogeneous mixture in which an insoluble solid is dispersed in a liquid; the solid particles are greater than 1,000 nm and generally large enough to be seen under a microscope

B. Liquid solutions

1. Substances vary in their capacity for dissolving in a given solvent; the degree to which a solute will dissolve in a given solvent is called its **solubility** with respect to that solvent

a. If large amounts of solute can be dissolved, the substance is highly soluble
   b. If only small amounts can be dissolved, the substance is slightly soluble
   c. If the intended solute will not dissolve, the substance is *insoluble*
   d. If a liquid solute dissolves readily in a liquid solvent, the two liquids are *miscible* (for example, milk and water mix together readily and therefore are miscible)
   e. Liquids that do not dissolve in each other are *immiscible* (for example, salad oil does not dissolve at all in vinegar)
2. The solubility of a solid solute in a liquid solvent varies with the temperature
   a. In most cases, the solute becomes more soluble as the temperature of the solvent increases
   b. For example, in making fudge, large amounts of sugar are mixed with small amounts of milk; initially, all of the sugar will not dissolve in the milk; but, when the mixture is heated to boiling, all of the sugar dissolves
3. *Dilute* solutions are those with a relatively low concentration of solute in solvent; solutions containing the maximum dissolvable amount of solute are called *saturated* solutions; solutions that contain more than the maximum dissolvable solute are *supersaturated*
   a. Saturated solutions exist in a state of equilibrium; the same number of ions or molecules are leaving the solutions as are dissolving
   b. When the equilibrium is disturbed (for example, by changing the temperature), the solute may come out of solution
4. The solubility of a gaseous solute in a liquid solvent depends on the pressure exerted on the gas
   a. For example, a can containing a soft drink exerts pressure on the solution of gas and liquid that makes up the drink; opening the can releases the pressure exerted by the can and allows the escaping gas (fizz) to escape from the liquid
   b. When the applied pressure increases, more gas dissolves in the liquid; conversely, when the applied pressure decreases, less gas remains in the liquid (this principle is known as *Henry's law*)
5. Scientists have devised various ways of calculating the exact concentrations of reactants in solutions; the most common methods involve measuring the molarity, molality, normality, and percent solution (these are discussed in more detail in Chapter 7, "Solutions")
6. Some physical properties of a solution vary with the number of solute particles dissolved in a given mass of solvent and do not depend on the chemical nature of the substances; these properties are called *colligative properties*
   a. The colligative properties of a liquid solution are its freezing point, boiling point, vapor pressure, and osmotic pressure
   b. In calculating how much a colligative property changes with the addition of a given solute, scientists must know how many particles each solute molecule yields in solution (for example, one mole of dissolved NaCl yields two moles of ions; one mole of K_2SO_4 yields three moles of ions [two moles of K^+ and one mole of SO_4^{-2}])
7. The addition of solute to a solution causes *freezing point depression;* for each

mole of particles dissolved in 1,000 grams of water, the freezing point is reduced 1.86° C
8. The addition of solute to a solution also causes *boiling point elevation;* for each mole of particles dissolved in 1,000 grams of water, the boiling point increases 0.52° C
9. The addition of solute to a solution reduces its vapor pressure by an amount proportional to the number of moles of nonvolatile solute present
10. The addition of solute to a solution increases its osmotic pressure
    a. *Osmosis* is the movement of water through a semipermeable membrane; if the membrane separates two solutions with different concentrations of solute, the water molecules tend to flow through the membrane and into the solution with the highest concentration of solute; the water will continue to flow toward the more concentrated side until the solute concentration on both sides of the membrane is equal
    b. *Osmotic pressure* is the amount of pressure necessary to stop osmosis; as the concentration of solute increases, osmotic pressure increases
11. If a solvent in a solution is volatile (easily forms a gas) and the solute is nonvolatile (rarely forms a gas), scientists can recover the solute by boiling away the solvent; *distillation* is the process of boiling away a volatile solvent to recover the solute
12. *Fractional distillation* is a process used to collect and separate two or more volatile components of a solution; for example, a solution in which liquid A (with a boiling point of 110° C) and liquid B (with a boiling point of 150° C) are dissolved in liquid C (with a boiling point of 190° C) can be separated into pure A, B, and C through fractional distillation
    a. In the solution, each component has a characteristic vapor pressure and boiling point
    b. As heat is applied to the solution, the component with the lowest boiling point (liquid A here) will vaporize first and leave the solution as a gas
    c. A cooling chamber collects the gas and condenses it back into pure liquid A
    d. Continued heating of the solution will allow separation of component B in a similar manner, leaving pure C behind
13. An *electrolyte solution* is a water solution (also called an *aqueous solution*) that is a better electrical conductor than pure water alone
    a. An *electrolyte* is a substance that dissolves in water to form ions; the ions conduct electricity through the water
        (1) Strong electrolytes are virtually totally ionic in water solution
        (2) Weak electrolytes are polar covalent substances that incompletely dissociate into charged particles in water; some of the dissolved molecules dissociate into charged ions, but most remain as whole molecules
    b. A *nonelectrolyte* is a substance that dissolves in water as whole, uncharged molecules; a nonelectrolyte does not enhance the flow of electricity through water

Types of Colloids

Here are examples of some common colloid mixtures.

TYPE	COMMON NAME	EXAMPLE
Gas in liquid	Foam	Whipped cream
Gas in solid	None	Marshmallows, Styrofoam
Liquid in gas	Aerosol	Fog, spray deodorant
Liquid in liquid	Emulsion	Mayonnaise, milk
Liquid in solid	None	Cheese, jellies
Solid in gas	None	Smoke
Solid in liquid	Sol	Paint, ink
Solid in solid	None	Colored glass

C. **Colloids**
   1. A colloid is a type of solution in which the particles are not dissolved but are dispersed throughout the solvent (or medium) and held in suspension
   2. Although many colloids are composed of particles in a liquid medium, various other types of colloids are in either a gaseous or solid medium (see *Types of Colloids*)
   3. Colloids exhibit the *Tyndall effect,* which is the ability to scatter and reflect light rays; light directed toward a colloid becomes a visible, defined beam as it passes through the colloid
   4. Colloid particles demonstrate *Brownian movement,* an erratic motion produced by the collision of larger colloid particles with the relatively small molecules found in the medium

D. **Suspensions**
   1. A suspension is a heterogeneous mixture of relatively large particles in a medium; in this type of mixture, the particles eventually settle out of the medium because of their large size or insolubility in the medium
   2. Suspensions may appear to be colloidal in nature; but because of the size of the suspended particles, settling (also called **precipitation**) will eventually occur
   3. After precipitation of the suspended particles, the suspension will cease to exist; instead two distinct layers appear—the clear medium and the precipitated particles

Study Activities

1. Devise an experiment showing the relationship of viscosity to temperature.
2. Demonstrate surface tension using a bowl of water and a straight pin; explain what happens when dish detergent is mixed with the water.
3. Explain colligative properties and outline the effects on each when solute is added to a solution.
4. Explain what happens when table salt is added to water during cooking.
5. Outline the steps for fractional distillation and give an example of its practical application.
6. Explain the Tyndall effect and Brownian motion in colloids.
7. Identify the differences between strong and weak electrolytes.
8. Identify the colligative properties of solutions.

10

Gases

Objectives

After studying this chapter, the reader should be able to:
- Identify the five characteristics of a gas.
- Describe how the kinetic molecular theory of gases explains gas pressure.
- State the three laws that combine to form the ideal gas law.
- Explain how the Kelvin temperature scale is used.
- State the three laws that combine to form the general gas law.

I. Properties of Gases

A. General information
1. A *gas* is a compound or element whose molecules or atoms are far apart and have much freedom of movement
2. Under normal conditions of temperature and pressure, both elements and compounds may exist as a gas
3. The *noble gases* (all of the elements in the periodic table Group 0) exist as monoatomic gases; these gases are unique in that they fail to react with any other elements under ordinary conditions
4. Generally, gases are characterized by their high compressibility, high expansivity, indefinite shape, low density, and quick diffusion
    a. *Compressibility* is a measure of how much a substance decreases in volume as a result of an increase in applied pressure; the gaseous state is the most compressible state of matter; gases are 100,000 times more compressible than solids
    b. **Expansivity** is a measure of the extent a substance will increase in volume as a result of a decrease in applied pressure; the gaseous state is the most expansive state of matter
    c. Because gases have no intermolecular forces of attraction, gas molecules or atoms have no definite shape and expand to fill all parts of any closed container until the molecules or atoms are equally distributed throughout the entire container
    d. Gases have the lowest density of all forms of matter; the densities of gases are about $1/1{,}000$ the density of any liquid or solid
    e. **Diffusion** is the spontaneous and even mixing of a gas with other gases; gases are infinitely miscible and form solutions with other gases quickly and easily

(1) In an uneven mixture of two or more gases, the gases will diffuse through random motion from areas of high concentration to areas of low concentration until they are evenly and randomly mixed throughout the container
(2) The diffusion of a particular respiratory gas, such as carbon dioxide or oxygen, depends on the partial pressure of that gas in the blood and in the lungs
(3) In human physiology, respiratory gases diffuse from areas of high concentration to areas of low concentration until they achieve equilibrium; for example, oxygen diffuses from the lung (high concentration in the pulmonary alveoli) into the blood stream (low concentration)
(4) A common test performed in clinical chemistry laboratories is arterial blood gas analysis, a test that measures the concentration of both oxygen and carbon dioxide in the blood

B. Kinetic molecular theory of gases

1. Gases differ from other forms of matter because their molecules have extremely low intermolecular forces of attraction and high kinetic energy
2. The *kinetic molecular theory of gases* postulates that gas molecules are extremely far apart and in constant, random motion
3. The moving gas molecules undergo frequent collisions with each other and with the walls of the container; these molecular collisions exert pressure or force on the container walls
4. The kinetic energy of gaseous molecules increases with increasing temperature; at any given temperature, the average kinetic energy of molecules of all gases is the same
5. Unlike solid or liquid molecules, gas molecules have no intermolecular forces of attraction or repulsion
6. The kinetic molecular theory not only explains the properties of gases, but also is the basis for all the gas laws
7. Six gas laws—Boyle's law, Charles' law, Gay-Lussac's law, the general gas law, the ideal gas law, and Avogadro's law—describe the physical nature of gases
8. Two gas laws—Dalton's law of partial pressures and Graham's law of effusion—explain the chemical interaction of gases

C. Gas pressure

1. In a closed container, gas exerts a measurable pressure on the walls of the container; for a given amount of gas in a closed container, the pressure increases as the temperature of the gas increases
2. Pressure is force per unit of area; any exertion of force applied to a definite area can be expressed mathematically as pressure;

$$P \text{ (pressure)} = \frac{F \text{ (force)}}{A \text{ (area)}}$$

3. Scientists use different units of pressure to measure certain gases; in many cases, the unit used depends on the equipment and on the conventional measurement standards

> **Measuring Gas Pressure**
>
> When measuring gas pressure, the units may vary depending on the specific gas and on the conventional equipment used. However, all units can be converted using the standards below.
>
> 1 atmosphere (atm) =
> 760 millimeters of mercury (mm Hg) =
> 760.0 Torricelli units (torr) =
> 14.7 pounds per square inch (psi)

      a. The air pressure of an automobile tire may be 35 psi (35 pounds per square inch); because **weight** is the force with which an object is pulled downward by gravity, the pound is a unit of force, and pounds per square inch is unit of pressure
      b. In weather analysis, the pressure of the surrounding air on the earth (the barometric pressure) is measured in inches; the inches refer to the height of a mercury column in a barometer; scientists use the same barometer, but measure the height of the mercury column in millimeters
4. The traditional instruments for measuring pressure in chemistry and in health care are the barometer and the manometer
      a. The barometer is a mercury-filled tube invented by the seventeenth-century physicist Torricelli; the height of the mercury in the tube directly reflects the surrounding air pressure
      b. The manometer is similar to the barometer, but it is used to measure the pressure of a closed sample of a gas
5. The traditional unit of pressure used in chemistry and health care is millimeters of mercury (abbreviated mm Hg)
      a. This unit arose from Torricelli's column of mercury; the original tube was approximately 850 mm long
      b. The distance in millimeters the mercury rises in the tube reflects the atmospheric pressure; 1 mm Hg is the atmospheric pressure sufficient to support a column of mercury 1 mm in height
      c. The average pressure of the earth's atmosphere at sea level and at 0° C supports a column of mercury to a height of 760 mm Hg; one *atmosphere* is the pressure necessary to support 760 mm Hg; 1 atm = 760 mm Hg
6. Other units used to measure gas pressure include torrs and pounds per square inch (see *Measuring Gas Pressure*)
      a. A torr (named in honor of Torricelli) is equal to $1/760$ atmosphere; 1 torr = 1 mm Hg
      b. In the English System, pressure is measured in pounds per square inch; 1 atm = 14.7 psi

D. **Gas density**
    1. The density of a gas is its mass per unit volume, usually expressed as grams per liter (g/L); the density of a gas is directly proportional to its molar weight
    2. When a gas is kept at a constant volume and temperature, its density will increase as its pressure increases; when a gas is at a constant pressure, its density will decrease as its temperature increases

Applying Boyle's Law

If you have 5 liters of a gas at 1 atmosphere of pressure and 0° C, what will be the pressure if the gas is compressed at the same temperature to 2 liters?

According to Boyle's law: PV = C; if we know P and V, we can calculate the constant (C).

Step 1. 1 atm × 5 liters = 5 L-atm
C = 5 L-atm

Using the constant, calculate the unknown pressure at the new volume:

Step 2. PV = C
P × 2 L = 5 L-atm
P = 5 L-atm/2 L
P = 2.5 atm

The pressure of the gas at the new volume will be 2.5 atmospheres.

  3. Because gas density varies with temperature and pressure, scientific tables list gas densities as measured at a *standard temperature* and *pressure* (also called *STP*)
     a. Standard temperature is 0° C
     b. Standard pressure is 1 atmosphere (760 mm Hg)
  4. Because the molecules of a gas are separated by large distances, the density of a gas is extremely low under ordinary atmospheric conditions

II. Gas Laws

A. General information
  1. Gas laws are mathematical equations that describe and predict the behavior of gases
  2. The properties of gases that are of critical interest in the gas laws include pressure (P), volume (V), absolute temperature (T), and number of moles of gas (n)

B. Boyle's law
  1. When a gas is compressed to a smaller volume, the molecules have less volume in which to move and therefore strike the surface of the container more frequently
     a. Thus, a decrease in the volume will cause an increase in the pressure of the gas
     b. This relationship between the pressure and the volume of a gas was described in the seventeenth century by British physicist Robert Boyle
  2. *Boyle's law* states that a gas at a constant temperature will have a volume that varies inversely with its pressure (see *Applying Boyle's Law*)
  3. If the volume of a gas increases, the pressure will decrease and vice versa
  4. Mathematically, the pressure of a gas at a constant temperature will equal a constant (C) divided by its volume; therefore, according to Boyle's law, P = C/V

> **Applying Charles' Law**
>
> If you have 100 liters of gas at 200° K and a pressure of 730 mm Hg, what will the volume of gas be at 300° K at the same pressure?
>
> Step 1. According to Charles' law, $V/T = k$
> $$(100\ L)/(200°\ K) = k$$
> $$0.500\ L/°K = k$$
>
> Step 2. Again using Charles' law, $V/(300°\ K) = 0.500\ L/°K$
> $$V = 0.500\ L/°K \times 300°\ K$$
> $$V = 150\ L$$
>
> The volume of the gas at the new temperature will be 150 liters.

5. This relationship also can be converted; at a constant temperature, the pressure multiplied by the volume will equal the constant (PV = C)
6. Another more useful variation of Boyle's law is in the equation $P_1V_1 = P_2V_2$

C. Charles' law and Gay-Lussac's law
  1. The eighteenth-century French scientist Jacques Charles described the relationship between the volume of gas and its temperature
  2. Charles found that the volume of a gas at a constant pressure varies directly with the temperature
     a. **Charles' law** states that at a constant pressure, the volume of a gas will equal its temperature multiplied by a constant; $V = kT$, where k is a constant that remains the same as long as the pressure of the gas is constant (see *Applying Charles' Law*)
     b. Other ways of writing Charles' law are as follows:
        $V/T = k$ and $V_1/T_1 = V_2/T_2$
  3. The work of Charles was extended by Joseph Gay-Lussac in 1808
     a. **Gay-Lussac's law** states that a gas at a constant volume has a pressure equal to its temperature multiplied by a constant; $P = kT$, where k is a constant as long as the volume of the gas is constant
     b. Other ways of writing Gay-Lussac's law are:
        $P/T = k$ and $P_1/T_1 = P_2/T_2$
  4. Both Charles' Law and Gay-Lussac's Law depend on use of the Kelvin temperature scale to measure gas temperature; this scale is different than Celsius and Fahrenheit scales; it rests on the assumption that gas may be theoretically cooled to a minimum temperature at which it would have zero volume, zero pressure, and no molecular motion
     a. This temperature is defined as absolute zero
     b. Using a simple technique of graphing temperature versus pressure, scientists calculate that the temperature of absolute zero is actually $-273°$ C
     c. To perform calculations using either Charles' law or Gay-Lussac's law, scientists must express the gas temperature in Kelvin degrees (°K)
     d. Because $0°\ K = -273°\ C$, then $273°\ K = 0°\ C$
     e. To convert any Celsius temperature to Kelvin degrees, simply add 273 to the Celsius temperature: $°K = °C + 273$

Applying the General Gas Law

If 1 liter of gas at STP is heated to 100° C and compressed to 3 atmospheres pressure, what will its new volume be?

Step 1. $P_1V_1/T_1 = P_2V_2/V_2$

Step 2. $P_1 = 1$ atm (because STP indicates 1 atm pressure)
$T_1 = 0°$ C (because STP indicates 0° C)
$T_1 = 273°$ K (using the equation °K = °C + 273)
$V_1 = 1$ L

$P_2 = 3$ atm
$T_2 = 100°$ C
$T_2 = 373°$ K
$V_2 =$ Unknown value

Step 3. Inserting the known values into the equation in Step 1, we can calculate V_2.

(1 atm)(1 L)/273° K = (3 atm)(V_2)/373° K

Solving the equation algebraically for V_2, we get:

Step 4. (1 atm)(1 L)(373° K)/(3 atm)(273° K) = V_2
0.46 L = V_2

The volume of the gas under the new conditions would be 0.46 liter.

D. **General gas law**
 1. The ***general gas law*** provides a mathematical equation for describing the behavior of gases: $P_1V_1/T_1 = P_2V_2/T_2$
 2. The general gas law—a simple algebraic combination of Boyle's law, Charles' law, and Gay-Lussac's law—permits scientists to predict the behavior of gases under various conditions of pressure, temperature, and volume
 3. All temperature measurements used in the general gas law must be converted to the Kelvin scale to ensure the accuracy of the calculations (see *Applying the General Gas Law*)

E. **Avogadro's law**
 1. Proposed by nineteenth-century Italian physicist Amadeo Avogadro, ***Avogadro's law*** states that equal volumes of all gases at the same temperature and pressure contain the same number of molecules
    a. Avogadro's law means that at the same temperature and pressure, one liter of hydrogen gas will contain the same number of gas molecules as one liter of oxygen or any other gas
    b. This law also means that at the same temperature and pressure, two liters of gas will contain twice as many gas molecules as one liter of gas
    c. Avogadro's law explains that at constant temperature and pressure, the volume of a gas is directly proportional to the number of moles
 2. Avogadro's law is mathematically expressed by the formula $V = kn$, where *k* is a constant and *n* is the number of moles of gas

Applying the Ideal Gas Law

Calculate the density of ammonia (NH_3) in grams per liter (g/L) at 752 mm Hg and 55° C.

Step 1. According to the ideal gas law: PV = nRT.

Step 2. We know that the number of moles of a gas (n) is equal to the mass of the gas in grams (m) divided by its molar weight in grams (M): n = m/M.

Step 3. Substituting into the equation in Step 1 above, we get: PV = mRT/M

Step 4. m = PVM/RT

Step 5. We know that density (d) is equal to mass in grams (m) per volume (V):
d = m/V
m = dV

Step 6. Substituting the value of m above into the equation in Step 4, we get:
dV = PVM/RT

Step 7. Simplifying the equation in Step 6, we get:
d = PM/RT, where d = gas density and M equals molar weight

Step 8. Using the equation in Step 7, we can calculate the density of ammonia under the given conditions:

P = 752 mm Hg
M = 17.03 g/mole (the molar weight of ammonia)
R = 62.4 (L)(mm Hg)/(mole)(°K) one of the values for the ideal gas constant
T = 328° K (the temperature 55° C expressed in °K)

$$d = \frac{752 \text{ mm Hg} \times 17.03 \text{ g/mole}}{62.4 \text{ (L)(mm Hg)/(mole)(°K)} \times 328° \text{ K}}$$

d = 0.63 g/L

The density of ammonia at 752 mm Hg and 55° C is 0.63 g/L.

F. Ideal gas law
1. The *ideal gas law* provides a mathematical equation for describing the behavior of gases: PV = nRT (see *Applying the Ideal Gas Law*)
    a. In the ideal gas law, *n* is the number of moles of gas present, *P* is the gas pressure, *V* is the gas volume, and *T* is the gas temperature in °K
    b. *R*, the universal gas constant, is similar to other natural constants, such as the speed of light or absolute zero
        (1) *R* may be expressed in numerous ways depending on the units needed to perform a calculation
        (2) R = 0.821 (L)(atm)/(mole)(°K) *and* R = 62.4 (L)(mm Hg)/(mole)(°K)
    c. The value of R used depends on the units of measurement needed to complete the calculation
2. The ideal gas law results from the algebraic combination of Boyle's law, Charles' law, and Avogadro's law

Applying Dalton's Law of Partial Pressures

A pressurized cylinder contains a mixture of nitrogen and oxygen in the ratio of 3 volumes of nitrogen to 1 volume of oxygen. The total pressure of the gases is 5 atmospheres. What is the pressure of the oxygen in the cylinder?

Step 1. According to Dalton's law of partial pressures:
$$P_{total} = P_{nitrogen} + P_{oxygen}$$

Step 2. According to Avogadro's law, equal volumes of gases at the same temperature and pressure contain equal numbers of molecules; therefore, the number of molecules of nitrogen present is three times the number of molecules of oxygen present. Because every gas behaves the same, the pressure exerted by the nitrogen is three times the pressure exerted by the oxygen.

Step 3. $P_{nitrogen} = 3\ P_{oxygen}$

Step 4. $P_{total} = 3\ P_{oxygen} + P_{oxygen}$

Step 5. $P_{total} = 4\ P_{oxygen}$

Step 6. $5.00\ atm = 4\ P_{oxygen}$

Step 7. $1.25\ atm = P_{oxygen}$

The pressure of the oxygen in this situation is 1.25 atmospheres.

G. **Dalton's law of partial pressures**
1. **Dalton's law of partial pressures** is named after the English scientist John Dalton, who noted that in a mixture of gases that do not react chemically with each other, the total pressure exerted by the mixture is the sum of the partial pressures of each of the gases
2. This can be mathematically stated as follows:
   $P_{total} = P_1 + P_2 + P_3 + P_4 + \ldots$
3. The **partial pressure** of a gas is the pressure that a gas in a mixture would exert if it were present alone under the same conditions
4. According to Dalton's law of partial pressures, the pressure a gas exerts in a mixture is the same as it would be if the gas were alone in the container under the same conditions (see *Applying Dalton's Law of Partial Pressures*)
5. In the laboratory, experimenters frequently collect the gas from a chemical reaction by displacing a volume of water in a closed container; at the end of the reaction, the total gas pressure in the closed container is the sum of the partial pressure of the collected gas and the partial pressure of water vapor
6. The inclusion of the partial pressure of water vapor to the total gas pressure is particularly significant in understanding the physiologic behavior of gases in the human body and body fluids

H. **Graham's law of effusion**
1. *Effusion* is the movement of a gas through a small opening to a region of lower pressure

Gas Laws

Applying Graham's Law of Effusion

What is the molecular weight of gas X if it effuses 0.876 times as rapidly as $N_2(g)$?

Step 1. $\dfrac{r_1}{r_2} = \sqrt{\dfrac{M_2}{M_1}}$ r_1 = rate of effusion of gas X
M_1 = molecular weight of gas X
r_2 = rate of effusion of nitrogen
M_2 = molecular weight of nitrogen

Step 2. $r_1 = 0.876 r_2$

Step 3. $M_2 = 28.02$ (the molecular weight of N_2)

Step 4. $\dfrac{0.876 r_2}{r_2} = \sqrt{\dfrac{28.02}{M_1}}$ or $0.876 = \sqrt{\dfrac{28.02}{M_1}}$

Step 5. Square both sides. $0.7674 = \dfrac{28.02}{M_1}$

Step 6. Solve for M_1. $M_1 = 36.5$

The molecular weight of gas X is 36.5 amu.

---

2. **Graham's law of effusion** concerns the movement of gases at varying rates, when temperature and pressure are constant; the law states that a gas effuses at a rate inversely proportional to the square root of its density, if pressure and temperature are constant (see *Applying Graham's Law of Effusion*)
3. Another form of Graham's law states that the effusion rate of a gas is inversely proportional to the square root of its molecular weight; $r = k/\sqrt{M}$, where r is the rate of effusion and M is the moleclular weight of the gas
4. Applying simple algebraic formulas, Graham's law can be stated as follows:

$$\dfrac{r_1}{r_2} = \sqrt{\dfrac{M_2}{M_1}}$$

I. **Henry's law**
   1. **Henry's law** defines the quantitative relationship between the solubility of a gas in a liquid and its pressure
   2. Henry's Law states that at a given temperature, the solubility of a gas in a liquid is directly proportional to the pressure of the gas over the liquid
   3. This relationship can be stated mathematically by the following formula: $C_1 P_2 = C_2 P_1$, where C is the concentration of the gas and P is the applied pressure (see *Applying Henry's Law,* page 70)

Applying Henry's Law

The solubility of pure nitrogen gas in water at 25°C and 1 atm is 6.8×10^{-4} mol/L. What is the concentration of nitrogen dissolved in water under atmospheric conditions? Remember that under atmospheric conditions, the partial pressure of nitrogen gas in the atmosphere is 0.78 atm.

According to Henry's law:

Step 1. $P_1C_2 = P_2C_1$
$P_1 = 1$ atm
$P_2 = 0.78$ atm
$C_1 = 6.8 \times 10^{-4}$ mol/L
$C_2 =$ Unknown

Step 2. $C_2 = P_2C_1/P_1$

Step 3. $C_2 = \dfrac{(0.78 \text{ atm})(6.8 \times 10^{-4} \text{ mol/L})}{1 \text{ atm}}$
$C_2 = 5.3 \times 10^{-4}$ mol/L

The solubility of nitrogen in water at 25°C and atmospheric conditions is 5.3×10^{-4} moles per liter.

Study Activities

1. Describe a gas and compare its properties to those of solids and liquids.
2. Explain the kinetic molecular theory.
3. Describe what would happen to an air-filled balloon if it were heated, cooled, taken to the top of a mountain, and held on the ocean floor.
4. Explain the gas laws illustrated by the actions in Activity 3.

11

Radioactivity and Nuclear Reactions

Objectives

After studying this chapter, the student should be able to:
- Identify the three major forms of radioactivity.
- Define nuclide, radioisotope, half-life, and nuclear decay.
- Describe the characteristic properties of each type of radioactivity.
- List the units that measure radiation and explain their use.
- Differentiate nuclear fusion from nuclear fission.
- Cite at least three examples of applications of radioactivity in health care.

I. Radioactivity

A. General information
1. The study of radioactivity involves examining elements by looking at the structure of the atomic nucleus
   a. An atom is composed of a nucleus surrounded by a cloud of rapidly moving electrons
   b. The study of ordinary chemical reactions involves discovering how atoms combine with each other to share or exchange electrons
   c. The study of radioactivity involves discovering how atoms change from one element to another by changing the structure of the nucleus
2. An element may have several different nuclear structures and still be the same element chemically
   a. Atoms of the same element with the same atomic number but different mass numbers are called *nuclides* (or *isotopes*); in nature, elements occur as mixtures of nuclides
   b. Nuclides of the same element differ from each other only in the number of neutrons present in the nucleus; for example oxygen exists as three different nuclides: $^{16}_{8}O$, $^{17}_{8}O$, $^{18}_{8}O$; each of these oxygen atoms has eight protons, but the first atom has eight neutrons, the second has nine neutrons, and the third has ten neutrons
   c. **Radioisotopes**, also called *radionuclides*, are nuclides possessing unstable nuclei that spontaneously emit energy in the form of radiation
   d. The unstable nucleus, called the *parent*, decays or emits particles of radiation to produce a new nucleus, called the *daughter nucleus*, which in turn may decay until a stable daughter nucleus is produced

e. This final daughter nucleus may be a more stable form of the original element or it may be a totally different element
3. **Radioactivity** is the spontaneous emission of radiation by unstable atomic nuclei, resulting in the formation of a new element
   a. *Radiation* is energy that is radiated or transmitted in the form of rays, waves, or particles
   b. The most common forms of radiation given off by radioactive elements are alpha (α), beta (β), and gamma (γ) radiation

B. **Alpha radiation**
   1. **Alpha radiation** is the emission of alpha particles from the atomic nucleus
   2. Alpha particles contain two protons and two neutrons and are the same as a helium nucleus ($_2^4He$)
   3. Alpha particles have a charge of +2 and a mass of 4
   4. Alpha particles are relatively large compared to other radioactive particles; they move at $1/10$ of the speed of light and are stopped by clothing, paper, and skin
   5. As alpha particles travel, they tend to remove electrons from other atoms and leave behind a trail of electrons and positively charged particles; because these charged particles can penetrate skin, alpha radiation causes skin burns

C. **Beta radiation**
   1. **Beta radiation** is the emission of beta particles from an unstable atomic nucleus
   2. A beta particle is a high-energy electron that is ejected from the nucleus when a neutron becomes a proton
   3. Beta particles have a charge of -1 and a mass of essentially 0
   4. Beta particles are high-energy particles, moving at approximately $9/10$ the speed of light
   5. Because of their small size and great speed, beta particles can penetrate skin and cause severe burns

D. **Gamma radiation**
   1. **Gamma radiation** is the emission of **gamma rays** from the unstable atomic nucleus
   2. Gamma radiation is a naturally occurring, high-energy electromagnetic radiation similar to X-rays
   3. Gamma radiation has no mass and no electrical charge, but it has extremely high energy that travels at the speed of light
   4. Gamma radiation penetrates deeply into the body, causing severe damage; it even can pass through several feet of concrete
   5. Some elements emit only gamma radiation and no other forms of radiation; because gamma radiation has no mass or charge, gamma ray emission does not cause alter the atomic mass or the atomic number of the element

E. **Cosmic radiation**
   1. **Cosmic radiation** (or cosmic rays) are subnuclear particles in the environment that originate from the sun and outer space; it is the naturally occurring radiation that constantly bombards all the earth's inhabitants

Radionuclides

This chart provides information about some common radionuclides, including their characteristic half-lives and type of radiation emitted.

ELEMENT	ISOTOPE	HALF-LIFE	RADIATION EMITTED
Hydrogen	$^{3}_{1}H$	12 years	Beta
Carbon	$^{14}_{6}C$	5,730 years	Beta
Phosphorus	$^{32}_{15}P$	14 days	Beta
Technetium	$^{99}_{43}Tc$	6 hours	Gamma
Iodine	$^{131}_{53}I$	8 days	Beta and gamma
Polonium	$^{214}_{84}Po$	1.6×10^{-4} seconds	Alpha and gamma
Radium	$^{226}_{88}Ra$	1,600 years	Alpha and gamma

2. Composed primarily of high-speed protons, cosmic radiation also includes beta particles, gamma rays, and the nuclei of other elements
3. In the atmosphere, cosmic radiation interacts with gaseous molecules to produce other particles, such as electrons, protons, and neutrons

F. **Units of radiation**
   1. The measurable rate of decay by a radioactive substance depends on the amount of radioactive substance and on the identity of the radioactive element in the sample
   2. The *curie* (Ci) — the standard unit of measurement for radiation — is equal to 3.7 $\times$ 10^{10} disintegrations per second (dps); 1 Ci = 3.7 $\times$ 10^{10} dps
   3. Named for the French scientists Pierre and Marie Curie, who were the codiscovers of radium, one curie is the disintegration rate of one gram of radium-226
   4. Usually, the samples of radioactive isotopes used in medical diagnosis are much smaller than a curie; common units of measure include the millicurie (10^{-3} curies) and the microcurie (10^{-6} curies)
   5. In the International System of Units (the SI system), the unit describing the rate of decay is the becquerel (Bq), named for Antoine Becquerel, a French physicist who discovered radioactivity; 1 Bq = 1 dps

G. **Half-life**
   1. The *half-life* of a radionuclide is the time it takes for one-half of a given amount of radionuclide to decay (see *Radionuclides*)
      a. After one half-life, one-half of a sample of radionuclide will have decayed to form a new element and one-half of the original radionuclide will remain
      b. After two half-lives, one-fourth of the original sample will remain
      c. With each succeeding half-life, one-half of the previous amount of original isotope remains in the original form

2. Each isotope has a characteristic half-life that cannot be altered or changed; it is unaffected by chemical reactions with other elements or by temperature changes

II. Nuclear Reactions

A. General information
1. All radionuclides—both naturally occurring and man-made—release alpha, beta, or gamma radiation in a process called *nuclear decay*
2. Nuclear decay alters the nuclear structure of an element, resulting in the production of a new element that is entirely different from the old one
3. All nuclear reactions result in changing one element into an entirely different element by altering its nuclear structure
4. All nuclear reactions obey the laws of conservation of matter and conservation of energy; when such reactions are written out, the matter and charges must balance on both sides of the equation, just as in all equations involving chemical reactions
5. Some nuclear reactions produce energy; the three processes by which this occurs include nuclear fission, nuclear fusion, and nuclear transmutation

B. Nuclear decay
1. **Nuclear decay** is the spontaneous emission of radiation from the nucleus of an element and the simlutaneous change of that element into a different element (or elements) of lower atomic weight
2. In nature, the most stable elements have nuclei containing equal numbers of protons and neutrons; naturally stable nuclides of elements with lower atomic weights contain equal numbers of neutrons and protons
3. Nuclides of elements with higher atomic weights demonstrate greater variability in the number of neutrons and protons contained in their nuclei
   a. These elements are unstable and tend to decay or emit radioactivity
   b. The two factors affecting the tendency to radioactive decay are the ratio of neutrons to protons in the nucleus and the size of the nucleus
   c. Elements with large nuclei (high atomic weight) and a ratio of neutrons to protons that is greater than 1 are more likely to emit radiation
4. Every known isotope of every element with an atomic number greater than 83 is radioactive
5. In an attempt to gain greater stability, radioactive nuclei tend to decay until the nucleus has an equal number neutrons and protons
   a. Those nuclei with too many neutrons tend to emit a beta particle in order to assume a more stable form; emission of a beta particle converts a nuclear neutron to a proton
   b. Those nuclei with too few neutrons tend to emit a positron (positively charged electron) or to capture an electron and incorporate it into the nucleus to achieve stability; either of these events converts a nuclear proton to a neutron
6. Nuclear decay reactions can be shown as equations just as ordinary chemical reactions are shown as equations

Recording the Mass and Charge of Nuclear Radiation

This chart shows examples of how to record the mass and charge of alpha, beta, and gamma radiations.

TYPE OF RADIATION	SYMBOL	MASS (amu)	CHARGE
Alpha (α)	$^{4}_{2}He$ (Helium nucleus)	4	+2
Beta (β)	$^{0}_{-1}e$ (Electron)	0	−1
Gamma (γ)	$^{0}_{0}\gamma$ (Light)	0	0

7. In balancing nuclear reaction equations, the sum of the mass numbers on both sides of the equation must be equal and the sum of the charges on both sides of the equation must be equal
8. An easy way of tracking the mass numbers in the equation is to write the mass number of each reactant and product; to track the charge, write the atomic number (the number of protons in the nucleus) of each reactant and product and the charge of each radioactive particle given off in the reaction
   a. Write all radionuclides with the atomic mass number and atomic number shown just before the symbol for the element
   b. Write all decay particles with their mass and charge shown just before the symbol for the particle (see *Recording the Mass and Charge of Nuclear Radiation*)

C. Nuclear fission
1. **Nuclear fission** is the splitting of an atomic nucleus into two or lighter nuclei by bombardment with incoming high-speed particles, such as neutrons
2. When a heavy nucleus with a mass greater than 230 amu absorbs an extra neutron, the nucleus splits into two lighter nuclei (usually between 80 and 160 amu), producing more neutrons and heat energy
3. In a typical fission reaction, the number of neutrons released is greater than the number of neutrons needed to start the reaction
4. The fission of uranium can be written as:

   $^{235}_{92}U + ^{1}_{0}n \rightarrow ^{236}_{92}U \rightarrow ^{139}_{56}Ba + ^{94}_{36}Kr + 3^{1}_{0}n + $ Energy

   where *n* stands for a neutron (a particle with zero charge and mass of 1 amu)
5. As each new uranium atom is split, the number of neutrons available to split other uranium atoms is tripled
6. The result is a *chain reaction*, a reaction that continues until all the original element is used up
7. Nuclear fission begins at ordinary temperatures; the atomic bomb is a fission reaction operating on the principle of an unstoppable chain reaction that releases tremendous energy
8. Fission changes heavy elements to lower-weight elements that usually emit beta particles

D. Nuclear fusion
1. **Nuclear fusion,** also called a *thermonuclear reaction*, is the combining of two small nuclei to form a larger, more stable nucleus

2. A by-product of fusion is the production of large quantities of energy, usually heat
3. Fusion will occur only at extremely high temperatures—millions of degrees Celsius
4. The sun is a giant nuclear fusion reaction that combines hydrogen atoms to produce helium plus heat and light energy

E. **Nuclear transmutation**
   1. *Nuclear transmutation* is any process that results in the creation of a new nucleus, either by radioactive decay or by bombarding an existing element with neutrons, electrons, or other nuclei
      a. Nuclear fission (the splitting of the nucleus) and nuclear fusion (the combining of two nuclei) are types of nuclear transmutation reactions
      b. Scientists also may bombard an element with neutrons, protons, or the nucleus of another element
      c. This process may cause the nucleus of the original element to incorporate the incoming particles and enlarge
   2. The bombardment of an element with heavy particles, such as neutrons or protons, is used to synthesize new elements with atomic weights greater (heavier) than those of naturally occuring elements; for example, elements with atomic numbers 94 through 109 have been synthesized by this technique
   3. The bombardment of uranium-238 with nuclei of the hydrogen-2 isotope (called deuterium) produced the first new element—uranium-239—which decays to another new element, subsequently named neptunium-239a

III. Radiation Effects on Matter

A. **General information**
   1. All forms of radiation have the potential to react with matter
   2. The three primary forms of radiation—alpha, beta, and gamma—lose energy chiefly by interacting with electrons of atoms
   3. Alpha and beta particles forcibly eject electrons from atoms and molecules with which they collide, leaving a trail of charged particles in their paths; alpha and beta particles are called *ionizing radiation*
   4. Gamma rays are also forms of ionizing radiation, but because they are electromagnetic waves with no charge, they cause less ionization than either alpha or beta particles; however, gamma rays have the greatest penetrating power of the three types of radiation and the capacity to cause the most cellular damage to the human body
   5. Most devices that measure or detect radioactivity, such as Geiger counters or photographic plates, actually detect the ions caused by the radiation

B. **Radiation interaction with matter**
   1. Scientists measure the amount of radiation that interacts with matter in special units called *roentgens;* the roentgen primarily applies to gamma radiation, the most powerful and penetrating radiation

2. The *roentgen* (r) is the amount of X-rays or gamma radiation that will produce 2.1×10^9 units of ionic electrical charge on 1 cubic centimeter (cc) of dry air at standard temperature and pressure
3. The roentgen measures how many ions are created by radiation; it does not reveal how much radiation actually is absorbed by a human

C. **Radiation absorption by tissues**
   1. Scientists use two different units to measures the absorption of radiation by tissues: the rad and the rem
   2. The **rad** (*r*adiation *a*bsorbed *do*sage) is 100 erg of energy absorbed by 1 gram of tissue; 1 rad = 100 erg/1 g of tissue
      a. The rad is used to measure radiation absorption by any biological tissue
      b. In the SI system, 1 gray (Gy) = 100 rad
   3. The **rem** (*r*oentgen *e*quivalent *m*an) is the absorbed dose of radiation that produces the same biological effect as one rad of therapeutic X-rays
      a. The rem is used primarily to measure radiation absorption by the human body
      b. The rem is the product of the absorbed dose in rads multiplied by the relative biological effectiveness factor (RBE); rem = rad × RBE
      c. The RBE is a number value that depends on the effectiveness of different types of radiation
         (1) For beta particles and gamma rays, RBE = 1
         (2) For neutrons and alpha particles, RBE = 10; basically, a dose of one rad of alpha radiation is equivalent to 10 rem
      d. In the SI system, 1 sievert (SV) = 100 rem

D. **Radiation in health care**
   1. Radioactive materials are now used in standard procedures for medical diagnosis, disease treatment, and research
   2. Although exposure to X-rays and radionuclides carries risks, these materials also aid in the early detection, noninvasive diagnosis, and treatment of cancer and other diseases; in most cases, the benefits of exposure outweigh the risks
   3. Radioactivity is an essential part of many analytical procedures used to confirm a diagnosis
      a. Radionuclides emit radiation at a constant, predictable level; once the radionuclides are in the body, diagnostic detecting devices can measure their presence to a high degree of sensitivity and specificity
      b. When radioactive substances are injected into the body, detection devices can reveal their uptake by specific organs or tissues, their rate of circulation through the body, and their rate of elimination from the body
   4. Radionuclides can be used to measure minute amounts of important body chemicals, such as vitamins, hormones, or drug levels, which would be extremely difficult to measure by ordinary chemical analysis
   5. Radiation therapy, such as that used in cancer treatments, is commonly used for the therapeutic destruction of living tissues
   6. Most radioactive substances are used outside of the body, external to the organs and tissues; for example, cobalt-60, a beta and gamma emitter, is used for cancer therapy

78 Radioactivity and Nuclear Reactions

7. Internal implantation or administration of a radionuclide concentrates the effects of the radiation in a particular kind of cell, such as the bone or thyroid

Study Activities

1. List the half-lives of the following isotopes: $^{234}_{90}$Th, $^{131}_{53}$I, and $^{218}_{85}$At.
2. Describe the processes by which alpha, beta, and gamma radiation occur.
3. Explain how a nuclear accelerator works.
4. Explain how nuclear fission differs from nuclear fusion; cite examples of the actual and potential uses of both types of reactions.
5. Describe five uses of ionizing radiation in the medical field.

12

Acids, Bases, and Buffers

Objectives

After studying this chapter, the reader should be able to:
- Define and characterize an acid.
- Define and characterize a base.
- State the Arrhenius and Brønsted-Lowry models of acids and bases.
- Describe the relationship between the relative strength of an acid and the strength of its conjugate base.
- Describe the pH scale and how it is used.
- Define ionization constant and explain how it applies to weak acids or weak bases.
- Explain how buffers resist changes in pH from added acid or base.
- Explain how titration can be used to determine the concentration of an unknown acid or base.

I. Acid-Base Reactions

A. General information
1. An *acid* is a compound that ionizes in water to produce hydrogen (H^+) ions; it readily donates protons to other substances and, when dissolved in water, creates solutions that conduct electricity, taste sour, and turn litmus paper red
2. A *base* is a compound that ionizes in water to produce hydroxide (OH^-) ions; it readily accepts protons from other molecules and, when dissolved in water, creates solutions that conduct electricity, taste bitter, feel slippery, and turn litmus paper blue
3. Various models and theories explain the behavior of acids and bases, but the two most useful theories are the Arrhenius theory and the Brønsted-Lowry theory
    a. The *Arrhenius theory* (developed in 1884 by the Swedish chemist Svante Arrhenius) describes an acid as a substance that ionizes in water to produce H^+ ions, and a base as a substance that ionizes in water to produce OH^- ions
    b. The *Brønsted-Lowry theory* (developed independently in 1923 by the Danish chemist Johannes Brønsted and English chemist Thomas Lowry) describes an acid as a substance that readily donates protons to other molecules, and a base as a substance that readily accepts protons from other molecules

> **Forming Conjugate Acid-Base Pairs**
>
> The equation below exemplifies how conjugate acid-base pairs form.
>
> $$\text{HCl} + \text{H}_2\text{O} \rightarrow \text{H}_3\text{O}^+ + \text{Cl}^-$$
> (strong acid) (weak base) (strong conjugate acid) (weak conjugate base)

      c. The Arrhenius model is useful mainly for acids and bases in water solutions; the Brønsted-Lowry model is the most useful explanation for acid-base reactions in physiology
4. The Brønsted-Lowry model can be used to describe acids and bases in terms of their strength
      a. A strong acid completely or almost completely ionizes to donate all of its protons; a weak acid donates only some of its available protons
      b. A strong base has a strong attraction for protons; a weak base has a weak attraction for protons
5. When an acid donates a proton to another molecule, it forms a ***conjugate base;*** when a base accepts a proton from another molecule, it forms a ***conjugate acid***
      a. Strong acids dissociate to form weak conjugate bases; weak acids dissociate to form strong conjugate bases
      b. Weak bases accept protons to form strong conjugate acids; strong bases accept protons to form weak conjugate acids
6. The reacting acid or base and its ensuing conjugate counterpart are commonly referred to as a *conjugate acid-base pair* (see *Forming Conjugate Acid-Base Pairs*)
7. Conjugate acid-base pairs differ from each other by only a single proton (see *Common Acids and Their Conjugate Bases*)
8. Acids are named according to the number and types of atoms in the acid molecule
      a. *Binary acids* consist of hydrogen plus a nonmetal; they are named with the prefix *hydro-* plus the name of the second atom with *-ic* as a suffix; for example, hydrochloric acid (HCl) and hydrosulfuric acid (H_2S) are both binary acids
      b. *Ternary acids* contain hydrogen and oxygen, plus a third element—usually carbon, nitrogen, phosphorus, chlorine, bromine, or sulfur; they usually are named for the oxidation state of the third element, with the suffix *-ic* used for the lower oxidation state and *-ous* used for the higher oxidation state; for example, sulfuric acid (H_2SO_4) and sulfurous acid (H_2SO_3) are both ternary acids
      c. *Polyprotic acids* can donate more than one proton in a reaction
          (1) Phosphoric acid (H_3PO_4) can donate three protons to the reaction
          (2) Sulfuric acid (H_2SO_4) can donate two protons to the reaction

B. **Ionization of acids and bases**
    1. Strong acids and bases ionize or dissociate completely in water
      a. Strong acids ionize in water to yield aqueous H^+ ions and aqueous anions; for example, $HBr \rightarrow H^{+(aq)} + Br^{-(aq)}$

Common Acids and Their Conjugate Bases

This chart includes examples of both strong and weak acids along with their conjugate bases.

ACID	CHEMICAL REACTION	CONJUGATE BASE
STRONG ACIDS		
Hydrochloric acid (HCl)	$HCl \rightarrow H^+ + Cl^-$	Chloride ion (Cl^-)
Nitric acid (HNO_3)	$HNO_3 \rightarrow H^+ + NO_3^-$	Nitrate ion (NO_3^-)
WEAK ACIDS		
Ammonium ion (NH_4^+)	$NH_4^+ \rightarrow H^+ + NH_3$	Ammonia (NH_3)
Bicarbonate ion (HCO_3^-)	$HCO_3^- \rightarrow H^+ + CO_3^{-2}$	Carbonate ion (CO_3^{-2})

     b. Strong bases ionize in water to yield aqueous OH^- ions and aqueous cations; for example, $KOH \rightarrow OH^{-(aq)} + K^{+(aq)}$
2. Weak acids and bases will ionize in water only partially
     a. The dissociation reaction of acids and bases is reversible (it proceeds in both a forward and a reverse direction)
     b. For weak acids and bases at equlibrium, the concentration of the undissociated acid or base is greater than the concentration of the dissociated ions
3. *Percent ionization* is the percentage of the total concentration of an acid or base that is in ionic form (see *Calculating Percent Ionization,* page 82)
     a. The higher the percentage of ionization, the stronger the acid or base
     b. The lower the percentage of ionization the weaker the acid or base
4. The **ionization constant** of an acid (K_a) or base (K_b) is a number that indicates the degree to which an acid or base dissociates at equilibrium (see *Calculating the Ionization Constant,* page 83)
     a. The higher the ionization constant for an acid or base, the stronger the acid or base
     b. The lower the ionization constant, the weaker the acid or base
     c. Because extremely strong acids and bases dissociate completely in water, their dissociation constants are rarely calculated; dissociation constants normally are used only for predicting the behavior of weak acids and bases
     d. Values of K_a and K_b are experimentally determined; typical K_a values range from 1.8×10^{-5} for acetic acid to 1.9×10^{-1} for iodic acid

C. **Typical acid-base reactions**
    1. Four different acid-base reactions are possibile: a strong acid with a strong base, a weak acid with a strong base, a weak base with a strong acid, and a weak acid with a weak base
    2. When equal equivalents of a strong acid and a strong base react, a **salt** is formed and the resulting solution is neutral (neither acidic nor basic)
       a. For example, in the reaction $HCl + NaOH \rightarrow H_2O + NaCl$, the donated H^+ and OH^- ions combine to form H_2O and the other ions form the salt NaCl, which remains dissociated as Na^+ and Cl^- ions in the water

Calculating Percent Ionization

To calculate the percent ionization, divide the number of moles per liter of ionized acid or base in the solution at equilibrium by the total moles per liter of acid or base in the starting solution. Then, multiply the result by 100%. The stronger the acid or base, the higher the percent ionization.

$$\text{percent ionization} = \frac{\text{[moles of ionized acid or base at equilibrium]}}{\text{[total original moles of acid or base]}} \times 100\%$$

  b. The Na^+ ion has no tendency to accept or donate protons, so it has no effect on the acidity or basicity of the solution
  c. The Cl^- ion is an extremely weak base (according to Brønsted-Lowry theory) and has no tendency to accept a H^+ ion; consequently, it has no effect on the acidity or basicity of the solution
  d. The resulting solution is neutral — neither acidic nor basic
 3. When the equal equivalents of a weak acid and a strong base react, the resulting solution is basic
   a. A weak acid is only partially ionized in solution, and a strong base is completely ionized
   b. In a typical reaction involving a weak acid and a strong base, such as $CH_3COOH + NaOH \rightarrow CH_3COO^- + Na^+ + H_2O$, the weak acid (acetic acid) forms a stronger conjugate base (acetate ion)
   c. The acetate ion tends to react with water to form acetic acid and OH^- ions in the solution; the result is a basic solution; Na^+ ions are present as "spectator" ions
 4. When the equal equivalents of a weak base and a strong acid react, the resulting solution is acidic
   a. A weak base is partially ionized in solution; a strong acid is completely ionized in solution
   b. In a typical weak base with strong acid reaction (for example, $HNO_3 + NH_4OH \rightarrow NH_4^+ + NO_3^- + H_2O$), the weak base (ammonium hydroxide) forms a stronger conjugate acid (ammonium ion)
   c. The ammonium ion tends to react with water to form ammonium hydroxide and H^+ ions in the solution; the result is an acidic solution
 5. When a weak acid and a weak base react, the final solution may be acidic, basic, or neutral
   a. Both the weak acid and weak base ionize in solution only to the extent predicted by their ionization constants
   b. Whether or not the final solution is acidic or basic depends on the K_a and K_b of each of the reactants and on the initial concentration of each of the reactants

D. **Formation of salts**
 1. A ***salt*** is an ionic compound formed by the reaction between an acid and a base

> **Calculating the Ionization Constant**
>
> To calculate the ionization constant of an acid or base, multiply the molar concentrations of the ionized particles at equilibrium, then divide the result by the molar concentration of undissociated acid or base at equilibrium. The stronger the acid or base, the higher the dissociation constant (K).
>
> $$K = \frac{[\text{concentration of cations at equilibrium}] \times [\text{concentration of anions at equilibrium}]}{[\text{concentration of undissociated acid or base at equilibrium}]}$$

2. Aqueous acid-base reactions are generally characterized by the following equation: acid + base ⇌ water + salt
   a. A *neutralization reaction* is the reaction of acid with base to form salt plus water; this is shown by the reaction arrow to the right (→)
   b. The salt formed through the reaction of an acid with a base is a strong electrolyte that will completely dissociate in water
   c. Salt *hydrolysis* is the reverse reaction of salt with water to form acid plus base; this is shown by the reaction arrow pointing to the left (←)
   d. The solution resulting from salt hydrolysis can be either acidic or basic
      (1) The aqueous solution of a salt derived from a strong base and a weak acid is basic; for example, an aqueous solution of sodium acetate (the salt of acetic acid and sodium hydroxide) is basic
      (2) The aqueous solution of a salt derived from a strong acid and a weak base is acidic; for example, an aqeous solution of ammonium chloride (the salt of ammonia and hydrochloric acid) is acidic
3. Acid salts take their names from the acids from which they derived
   a. Binary acid salts are derived from binary acids; they take their name from the nonmetallic negative ion contained in the acid; their names always end in *-ide*
   b. For example, NaBr is sodium bromide; KCl is potassium chloride
4. Oxoacid salts contain polyatomic ions, such as SO_4^- and NO_3^-
   a. Oxoacid salts formed from an *-ic* acid have names that end in *-ate;* for example, a sulfate is a salt formed from sulfuric acid
   b. An oxoacid salt formed from an *-ous* acid have names that end in *-ite;* for example, a sulfite is a salt formed from sulfurous acid
   c. Polyatomic acid salts that still contain hydrogen atoms are named according to the number of hydrogen atoms present; for example, NaH_2PO_4 is sodium dihydrogen phosphate; Na_2HPO_4 is sodium hydrogen phosphate or disodium hydrogen phosphate

II. pH

A. General information
1. The terms *strong* and *weak* in reference to acids and bases indicate the extent of the ionization of the acid or base in solution; strong acids have a relatively high K_a; strong bases have a relatively high K_b

> **Calculating the Dissociation Constant of Water**
>
> Because the concentrations of H_3O^+ and OH^- for pure water at 25° C are always the same, you can calculate a dissociation constant for water similar to the the way you calculate a dissociation constant for an acid or a base.
>
> Step 1. $K = \dfrac{[H_3O^+][OH^-]}{[H_2O]}$
>
> Step 2. $K[H_2O] = [H_3O^+][OH^-]$
>
> Step 3. Because the concentration of water is always the same, assign K_w as the value of $K[H_2O]$.
>
> Step 4. $K_w = [H_3O^+][OH^-]$
>
> Step 5. $K_w = (1 \times 10^{-7})(1 \times 10^{-7})$
>
> Step 6. $K_w = 1 \times 10^{-14}$

   2. However, the strength of an acid or base in an aqueous solution also is reflected directly in the actual concentration of the H^+ or OH^- ions in the solution
   3. The **pH system** is based on actual concentration of H^+ ions in solution; it is related to the role of water in the ionization of acids and bases

B. The role of water ionization
   1. Water has the ability to dissolve both ionic compounds and highly polar molecules
   2. To a small extent, water also dissolves other water molecules to form H_3O^+ ions and OH^- ions
   3. At 25° C, one water molecule out of every 550 million will ionize to form a hydronium ion (H_3O^+) and hydroxide ion (OH^-); single H^+ ions never exist in water; they are always found as part of the hydronium ion (H_3O^+)
   4. This reaction of the ionization of water may be expressed in the following equation:
   $$H_2O + H_2O \rightarrow H_3O^+ + OH^-$$
   5. Scientists have measured the concentrations of H_3O^+ and OH^-; in pure water at 25° C, the concentrations of H_3O^+ and OH^- are each 1×10^{-7} moles
   6. When the ionization constant of water is calculated, the fixed value is 1×10^{-14} (see *Calculating the Dissociation Constant of Water*)

C. The pH scale
   1. The pH scale is simply a reflection of the actual concentration of the H_3O^+ ion in an aqueous solution; mathematically, $pH = -\log[H^+]$
   2. Because the concentration of H^+ in pure water is the same as the concentration of H_3O^+, the concentration of H^+ in pure water is 1×10^{-7}
      a. Pure water is neutral—neither acidic nor basic
      b. The pH of a neutral solution is 7 because $-\log[H^+] = -\log[10^{-7}] = 7$

The pH Scale: A Measure of Acid-Base Concentration

The pH scale reflects the concentration of hydronium ions (H_3O^+) in an aqueous solution. The scale below shows the relative concentration of acid (H^+) and base (OH^-) for each pH value on the scale.

pH	[H^+]	[OH^-]	RELATIVE ACIDITY OR BASICITY
0	10^0	10^{-14}	Strong acid
1	10^{-1}	10^{-13}	↓
2	10^{-2}	10^{-12}	↓
3	10^{-3}	10^{-11}	↓
4	10^{-4}	10^{-10}	↓
5	10^{-5}	10^{-9}	↓
6	10^{-6}	10^{-8}	Weak acid
7	10^{-7}	10^{-7}	Neutral
8	10^{-8}	10^{-6}	Weak base
9	10^{-9}	10^{-5}	↑
10	10^{-10}	10^{-4}	↑
11	10^{-11}	10^{-3}	↑
12	10^{-12}	10^{-2}	↑
13	10^{-13}	10^{-1}	↑
14	10^{-14}	10^0	Strong base

3. Because the lowest concentration that H^+ can have in water is 10^{14} and the highest concentration H^+ can have in water is 10^0, the pH scale runs from 0 to 14 (see *The pH Scale: A Measure of Acid-Base Concentration*)
   a. A pH between 0 and 7 is acidic
   b. A pH of 7.000... is exactly neutral
   c. A pH between 7 and 14 is acidic
4. In laboratory work, either a pH meter or indicator dyes give a measurement of pH (see *The pH of Common Fluids,* page 86, for examples of some commonly measured substances)
   a. A *pH meter* is an instrument that detects an electric current passing through a solution; the amount of current will change as the pH of the solution changes
   b. Chemical dyes, known as *acid-base indicators,* will change color at particular hydrogen ion concentrations; however, these are less accurate than a pH meter

D. Titration procedures
   1. **Titration** is a procedure that uses a neutralization reaction to determine the normality (the number of equivalents per liter of solution) of an unknown acid or base solution

The pH of Common Fluids

The pH values of some common fluids are listed below.

FLUID	pH
Stomach acid	1.0 to 2.0
Lemon juice	2.4
Vinegar	3.0
Orange juice	3.5
Urine	4.5 to 7.5
Saliva	6.4 to 6.9
Milk	6.5
Pure water	7.0
Blood	7.35 to 7.45
Intestinal secretions	7.7
Bile	7.8 to 8.8
Milk of magnesia	10.6
Household ammonia	11.5

    a. In titration, scientists measure the volume of a solution with known acid or base concentration (called the *titrant* or *standard solution)* required to completely neutralize another acid or base solution of unknown concentration

    b. Such measurements require the use of either an indicator dye or a pH meter to determine when neutralization occurs

2. The *equivalence point* is the volume of added titrant that contains exactly enough acid or base equivalents to completely neutralize the acid or base equivalents in the unknown solution
3. At the equivalence point, the pH of the unknown solution changes to exactly 7.0 and becomes completely neutralized
4. For a given reaction, an equivalent of acid is the amount of acid that will donate one mole of hydrogen ion; an equivalent of base is the amount of base that will neutralize one mole of hydrogen ion
5. This equivalent of reactive acid or base is measured in grams and is called the *gram-equivalent weight* or normality; mathematically, it equals the number of equivalents of solute per number of liters of solution
6. In a neutralization reaction, the number of equivalents of acid equals the number of equivalents of base; mathematically, this may be expressed by the formula $N_{acid} V_{acid} = N_{base} V_{base}$

III. Buffers

A. General information
1. A **buffer** is a solution that will resist sudden changes in pH when an acid or a base (or both) are added to it
2. Buffer systems consist of a weak acid and its conjugate base or a weak base and its conjugate acid—for example acetic acid ($HC_2H_3O_2$) and acetate ion ($C_2H_3O_2^-$)
   a. When H^+ ions are added to the acetic acid – acetate solution, the acetate ions react with the H^+ ions to form undissociated acetic acid, thus preventing the added H^+ ions from affecting the overall pH of the solution
   b. When OH^- ions are added to the solution, they react with the H^+ ions already dissociated from the acetic acid to form water; more acetic acid ions will then dissociate to restore the balance of H^+ originally present in the buffer
3. The most efficient buffers are made from equal concentrations of an acid and its conjugate base or equal concentrations of a base and its conjugate acid

B. Buffers in the human body
1. Buffers are particularly critical in living organisms; they protect the human body from acids produced by normal or abnormal cell metabolism
2. Buffers help prevent *acidosis,* an abnormal physiologic state that results when too much acid is produced or retained by the body; buffers also help prevent *alkalosis,* an abnormal physiologic state that results when too much acid is lost from the body or when too much base is ingested
3. Buffers are found in varying concentrations in all body fluids and within cells to protect against potentially harmful changes in pH
   a. The carbonic acid – bicarbonate ion buffer system maintains normal blood pH
   b. The dihydrogen phosphate ion – monohydrogen phosphate ion buffer system maintains normal pH of urine

Study Activities

1. Write the names and formulas of three common acids and three common bases.
2. List five acid-base reactions and identify their respective conjugate pairs.
3. Measure the pH of various household substances (such as coffee, lemon juice, and bleach), then plot their relative acidity or basicity on a pH scale.
4. Describe what occurs in a neutralization reaction; write four balanced equations demonstrating this type of reaction.
5. Explain the physiologic states of acidosis and alkalosis.

13

Metals and Metalloids

Objectives

After studying this chapter, the reader should be able to:
- Describe the principal chemical and physical properties common to most metals.
- Identify the three classes of metals.
- Compare and contrast the chemical properties of alkali metals and alkaline earth metals.
- Describe the chemical properties characteristic of transition metals.
- List the metals that are essential trace elements in human nutrition.
- Explain the differences between metals and metalloids.
- Describe the various uses of metals and metalloids in industry and biology.

I. The Properties of Metals

A. General information
1. All of the elements on the periodic table may be classified as **metals, metalloids, nonmetals,** or noble gases
2. About three-fourths of the elements on the periodic table may be classified as metals (see *Identifying Metals and Metalloids on the Periodic Table*)
3. Metals share certain physical characteristics
   a. All metals except mecury are solids at room temperature; mercury is liquid at room temperature
   b. All metals conduct heat and electricity easily
   c. All metals are malleable (can be shaped by hammering), ductile (can be drawn into a wire), and shiny
   d. Most metals have high melting points and are generally denser than nonmetals
4. In chemical reactions, metals lose electrons easily; they are easily oxidized to cations
5. The physical and chemical characteristics of metals result from the special type of bonding between the atoms in a solid state; these metallic bonds are simply a closely packed group of metal cations surrounded by a sea of constantly moving electrons

B. Classifying metals
1. Metals fall into three different classifications—alkali, alkaline earth, and transition—each having a different set of chemical properties

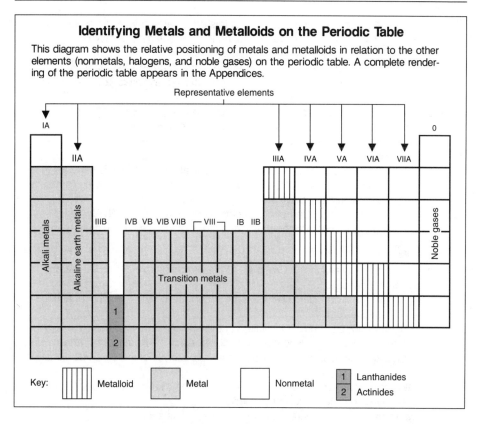

Identifying Metals and Metalloids on the Periodic Table

This diagram shows the relative positioning of metals and metalloids in relation to the other elements (nonmetals, halogens, and noble gases) on the periodic table. A complete rendering of the periodic table appears in the Appendices.

2. Alkali metals are good oxidizing and reducing agents
3. Alkaline earth metals are good oxidizing and reducing agents but do not react as readily as the alkali metals
4. Transition metals can assume several different oxidation states and thus may form two or more different compounds with a single different element

II. Metal Groups

A. General information
1. On the periodic table, all of the elements in Groups IA through IIB except hydrogen are metals
   a. Hydrogen is a gas and therefore does not fit in with the mostly solid metals
   b. However, hydrogen does react like a metal in some chemical reactions
2. Group IIIA contains four metals and one metalloid; Group IVA contains two metals and two metalloids; Group VA contains one metal and two metalloids; Group VIA contains two metalloids; and Group VIIA contains one metalloid

B. Group IA
1. Those elements found in Group IA of the periodic table are called *alkali metals*

a. The most abundant members of this group are sodium, potassium, and lithium, all of which can be found in a wide variety of minerals
b. Rubidium, cesium, and francium are less common elements
2. All alkali metals have one electron in the outer electron shell that is easily lost; because alkalis easily lose their outer electron, they make good reducing agents
3. Alkali metals react vigorously with water and oxygen; for example, sodium reacts spontaneously with water or air to form hydroxides or oxides
4. Alkali metals are never found in a metallic form in nature; rather, they usually exist as salts
5. Sodium and potassium are the most biologically significant alkali metals
   a. Both sodium and potassium are essential electrolytes, helping to maintain normal internal body stability (homeostasis) in man
   b. Sodium ions are the principal cations of extracellular fluids in man
   c. Potassium ions are the principal cations of intracellular fluids in man

C. **Group IIA**
   1. Those elements found in Group IIA of the periodic table are called *alkaline earth metals*
      a. The most abundant metals in the group are beryllium, calcium, and magnesium
      b. Strontium, barium, and radium are less common elements
   2. All alkaline earth metals contain two electrons in their outer shell; alkaline earth metals easily lose their outer electrons to become ions with a +2 oxidation state
   3. Alkaline earth metals are not as reactive as the alkali metals, but all will react slowly with water to form hydroxides
   4. Most alkaline earth metals are found in nature as salts
   5. Calcium and magnesium are biologically the most essential alkaline earth metals to man
      a. Calcium is part of the structure of bones and teeth and is critical for blood coagulation and nerve impulse transmission
      b. Magnesium is an important activator of body enzyme systems and also is crucial in regulating nerve function and muscle contraction

D. **Groups IIIB to IIB**
   1. Those elements found in Groups IIIB to IIB of the periodic table are called **transition metals**
   2. Important elements in this group include iron, copper, silver, and mercury, as well as many of the trace elements essential for biological function
   3. Transition metals have varying electron shell configurations
      a. The metals in Period 4—scandium through zinc—have electrons in the fourth shell but an incomplete third shell
      b. The metals in Period 5—yttrium through cadmium—have electrons in the fifth shell but an incomplete fourth shell
   4. Some transition metals—including iron, copper, and mercury—have more than one oxidation state and thus may form two or more compounds with the same element; for example iron forms several different compounds with oxygen, including ferrous oxide (FeO) and ferric oxide (Fe_2O_3)

5. Different oxidation states are possible because these elements may lose electrons from two different outermost electron shells
6. The electronegativity of transition metals increases from left to right across the periodic table; those metals with higher electronegativity are less reactive than the alkaline earth metals
7. Transition metals can form covalent compounds as well as ionic compounds
8. Most of the transition metals are found as salts; some transition metals, such as copper and gold, are occasionally found as pure elements in nature
9. A few of the transition metals are significant to humans as trace elements *(trace metals);* they are essential to health but normally are found in small quantities in animals and humans
    a. Copper is a trace metal necessary for the normal functioning of all living cells
    b. Iron is found in many important enzymes and electron carrier molecules; the most important biological function of iron is the transport of oxygen by iron-containing hemoglobin
    c. Zinc is essential for many enzyme systems, normal liver function, and the synthesis of deoxyribonucleic acid (DNA)
    d. Cobalt is an essential component of the vitamin B_{12} molecule
    e. Other trace element metals essential for human health include manganese, chromium, and nickel
10. Many transition metals can form chelates or coordination compounds with anions or with neutral molecules that have unshared electrons to donate
    a. A *chelate* is a complex molecule in which a metal ion or atom is bound at two or more points to another molecule, forming a ringlike structure
    b. Many chelates occur in nature, including heme (an iron chelate that is the oxygen-carrying portion of blood hemoglobin) and chlorophyll (a magnesium chelate that is the light-absorbing pigment of plant cells) (see *The Structure of a Heme Chelate,* page 92)

E. **Group IIIA**
   1. The most abundant metal of Group IIIA is aluminum; gallium, indium, and thallium are less commonly found members of this group
      a. Aluminum, a silvery white element, is the most abundant metal on the earth's crust (constituting about 8% of the crust)
      b. Aluminum has low density, malleability, and high tensile strength (it can withstand great stress without breaking)
      c. Gallium, indium, and thallium are widely distributed in nature but occur only in trace amounts
   2. All elements in Group IIIA are fairly reactive but will not form simple anions
      a. They mostly form covalent bonds with other elements
      b. They commonly assume an oxidation state of +3
   3. Of the metals in Group IIIA, only aluminum has uses in commerce and manufacturing
      a. Aluminum is used for aircraft construction, cans for food and liquids, and high-voltage transmission lines; it also is used as solid fuel or propellant for rockets
      b. Aluminum compounds also are used in water treatment plants for clarification

> **The Structure of a Heme Chelate**
> This diagram shows the molecular structure of a heme chelate.

      c. Aluminum has limited applications for humans; it is used in some antacids and deodorants; some research suggests that it may play a role in Alzheimer's disease

F. **Group IVA**
    1. The metals of this group are lead and tin
        a. Neither of these metals is abundant in nature
        b. Lead is an especially soft, malleable metal with a relatively low melting point
        c. Tin is a metal with a relatively low melting point and low tensile strength
    2. Group IVA metals usually form covalent bonds with other elements; they are fairly reactive and amphoteric (they can act as an acid or a base, depending on conditions)
    3. Group IVA metals have various commercial uses
        a. Lead is a useful metal for manufacturing and other industrial uses; for example, it is used in lead storage batteries and as a protective shield against radiation; however, it is biologically toxic
        b. Tin is used primarily to form alloys, such as bronze, solder, and pewter; it has no biological function

III. Metalloids

A. **General information**
    1. Metalloids include those elements that demonstrate variable characteristics of both metals and nonmetals
        a. Some metalloids are able to conduct electricity to a limited extent
        b. Other metalloids have relatively low melting points

c. As a group, metalloids demonstrate varying degrees of hardness
2. Metalloids include the following elements: boron (Group IIIA), silicon and germanium (Group IVA), arsenic and antimony (Group VA), tellurium and polonium (Group VIA), and astatine (Group VIIA)

B. **Boron**
1. Boron constitutes about 0.0003% of the earth's crust and occurs most commonly in nature as borates (the most familiar forms of boron include boric acid and neutralized boric acid [borax])
2. Boron is characterized by its low electrical conductivity, extreme hardness, brittleness, and high melting point
3. Used as a hardener in steel and aluminum alloys, boron is also used to make heat-resistant borosilicate glass (Pyrex)
4. Because of its acidic properties, boric acid is a suitable antiseptic for eye-wash and eye-care products

C. **Silicon**
1. Silicon is the second-most abundant element in the earth's crust, constituting about 28%
2. Silicon does not exist as a pure element in nature; rather, it is found primarily combined with oxygen as silicates (examples of silicates include asbestos and beryl; quartz is an example of silicon dioxide)
3. A semiconductor of electricity, silicon is extremely hard and has a high melting point
4. Most silicon compounds are formed by covalent bonding; they are fairly reactive and amphoteric (can act as either an acid or a base, depending on conditions)
5. Silicon is the major component of glass; because it is a semiconductor, it is used in transistors; it also is an important ingredient in steel because of its acid-resistant properties

D. **Germanium**
1. A rare element, germanium is hard, brittle, white, and metal-like in appearance
2. Germanium acts as a semiconductor of electricity and has a relatively high melting point
3. This element usually forms covalent bonds with other elements; it is fairly reactive and amphoteric
4. Because it is a good semiconductor, germanium is used in transistors and in integrated circuits; it has no known biological function

E. **Arsenic**
1. Arsenic can be found in various ores in nature; it constitutes about 0.0005% of the earth's crust
2. Arsenic exists in a soft, yellow, nonmetallic, and unstable form that is readily converted to a more stable gray, metallic form
3. In its more stable form, arsenic is comparatively soft and brittle and has a metallic luster; a conductor of electricity, arsenic has a low electronegativity and does not tend to form anions; most compounds are formed by covalent bonding

4. Extremely poisonous, arsenic is used in pesticides; however, recent research suggests that in animals, arsenic is an essential trace element required for adequate growth and reproduction

F. Antimony
1. Antimony, which is very similar to arsenic, exists as a native ore and can be found in the ores of copper, lead, silver, and mercury; it constitutes about 0.0005% of the earth's crust
2. It exists in a soft, yellow, nonmetallic, and unstable form that is readily converted to a more stable gray, metallic form
3. The more stable form is comparatively soft and brittle and has a metallic luster; a conductor of electricity, antimony has a relatively low electronegativity and does not tend to form anions; most compounds are formed by covalent bonding
4. An important constituent of metal alloys, antimony is added to lead to harden it for storage in batteries; it can be extremely poisonous

G. Tellurium
1. Tellurium is found in rare minerals and sulfur ores; it occurs naturally as metallic crystals
2. This silver-white element has a relatively high melting point and forms compounds primarily by covalent bonding
3. In oxidation states $+4$ and $+6$, tellurium is an effective oxidizing agent; it forms compounds similar to the way sulfur does; telluric acid is a stronger oxidizing agent than sulfuric acid
4. Tellurium is used to manufacture certain metal alloys; it also is used as an oxidizing agent in the synthesis of organic substances

H. Polonium
1. Polonium is produced only in the laboratory and results from the disintegration of radium
2. All polonium isotopes are radioactive
3. The most stable isotope, ^{210}Po, has a half-life of 138.7 days
4. The most metallic of all the elements in Group VIA, polonium appears to form cations in aqueous solutions
5. Polonium has no known applications or uses in commerce or human physiology

I. Astatine
1. Astatine, a member of the halogen family, was discovered through synthesis and radioactive decay
2. All astatine isotopes are radioactive; the most stable isotope, ^{210}At, has a half-life of 8.3 hours
3. Because of its nature, astatine is both difficult and expensive to study
4. Astatine has no known applications or uses in commerce or human physiology

Study Activities

1. Identify all of the metals and metalloids on the periodic table.
2. Write the chemical formulas of five different compounds containing iron and oxygen.
3. Using the Aufbau principle, write the electron configuration of two different alkali metals, two different alkaline earth metals, and one transition metal.
4. Describe the principle involved in metallic bonding.
5. List seven metals that are essential to human nutrition, and explain their biological significance.

14

Nonmetals and Noble Gases

Objectives

After studying this chapter, the reader should be able to:
- Characterize the physical and chemical properties of the elements classified as nonmetals.
- Describe the types of chemical bonds formed by several different nonmetals.
- Compare and contrast the chemical properties of the halogens and the inert gases.
- Identify and describe six nonmetals that are physiologically essential to humans.

I. Nonmetals

A. General information
1. All elements on the periodic table may be classed as metals, metalloids, nonmetals, or noble gases
2. Although nonmetals are a diverse group of elements, they share a few physical and chemical properties
   a. Nonmetals are nonshiny, have relatively low melting points and low densities, and, with the exception of carbon in the form of graphite, are poor conductors of electricity
   b. A common chemical property of nonmetals is their high electron affinity, which makes them tend to form ionic bonds or anions; most nonmetals also can form covalent bonds
   c. Some nonmetals, such as carbon, sulfur, and iodine, are solids; some, such as oxygen and nitrogen, are gases; only bromine is a liquid at room temperature
3. The *nonmetals* include hydrogen from Group IA; carbon from Group IVA; nitrogen and phosphorous from Group VA; oxygen, sulfur, and selenium from group VIA; and the halogens in group VIIA
4. The four most abundant nonmetals in the human body are carbon, hydrogen, oxygen, and nitrogen

B. Hydrogen
1. Although placed above the alkali metals (Group IA) in the periodic table, elemental hydrogen is a colorless, odorless, harmless gas, existing as H_2 molecules
   a. It is the lightest element and a potent reducing agent

- b. Hydrogen resembles the alkali metals because it can be oxidized to the H^+ ion; it also resembles the halogens because it can be reduced to the H^- (hydride) ion
- c. Hydrogen atoms commonly share electrons and easily form covalent bonds with other elements
2. Hydrogen is the most abundant element in the universe; it is found in outer space as a pure element and comprises most of the mass of the sun
3. On earth, hydrogen is found in compounds, such as water and all of the organic substances that make up the physical bodies of humans, animals, and plants
4. Hydrogen is highly significant to human physiology
    - a. Hydrogen constitutes about 63% of the total number of atoms found in the human body
    - b. Hydrogen is an important component of biological materials, such as proteins, carbohydrates, and lipids
5. Nonbiological uses of hydrogen include the hydrogenation of unsaturated fats to make margarine, synthesis of ammonia (fertilizers), and fuel for energy production (electricity); as fossil fuels diminish, hydrocarbon gas could become an important fuel source

C. Carbon
1. Carbon is the only nonmetal in group IVA; a solid in its elemental form, carbon may vary in color depending upon the impurities associated with it
    - a. Pure carbon has an extremely high melting point; carbon in the form of diamond is the hardest substance known
    - b. Carbon has four valence electrons; it almost always forms four strong, stable covalent bonds with other atoms, most commonly with hydrogen
    - c. The entire study of organic chemistry is the study of **hydrocarbons,** a large group of chemical compounds composed primarily of carbon and hydrogen
2. Carbon is essential for all living organisms; it is extremely important in human physiology
    - a. Carbon makes up about 9.5% of the total number of atoms in the human body
    - b. Carbon-containing compounds are the building blocks of life; the chemicals that make up the human body, such as proteins, fats, and lipids, are made almost entirely of carbon compounds
3. Nonliving sources of carbon include atmospheric carbon dioxde (CO_2), carbonate ions (CO_3^{-2}), limestone and chalk, complex hydrocarbon molecules in coal and petroleum, and diamonds

D. Nitrogen
1. Nitrogen is one of two nonmetals in group VA; elemental nitrogen is a colorless, odorless gas (N_2) that makes up about 78% of the earth's atmosphere
    - a. Nitrogen has five valence electrons and usually forms three covalent bonds to attain stability

b. Nitrogen can form strong bonds with carbon, oxygen, and hydrogen; its capacity to bond with these elements makes it an important element in organic chemistry
  2. Nitrogen is highly significant to human physiology
    a. Nitrogen makes up about 1.4% of the total number of atoms in the human body
    b. It is an essential component of proteins and the nucleic acids that make up deoxyribonucleic acid (DNA)
  3. People and animals cannot use atmospheric nitrogen (N_2) or soilborne nitrogen directly; atmospheric and soil nitrogen becomes available to people and animals through the nitrogen cycle
    a. Most plants take up nitrogen as nitrates from the soil; some plants can absorb nitrogen directly from the air through a unique process called *nitrogen fixation;* plants convert nitrogen taken from soil or air to proteins
    b. People and animals eat these plant proteins and thus incorporate nitrogen into their bodies
    c. Nitrogen returns to the soil as ammonia in the excretia of people and animals and through the decay of bodies after death
    d. Soil-dwelling microorganisms convert this nitrogen back to nitrates that can be used by plants, thus completing the nitrogen cycle
  4. Nitrogen also is found in commercial fertilizers; explosives; nylon; nitric acid, a major industrial inorganic acid; and nitrous oxide, a colorless gas used as an anesthetic for minor surgery

E. **Phosphorus**
  1. Phosphorus is one of two nonmetals in group VA; elemental phosphorus (P_4) exists in several forms, including white phosphorus, which spontaneously bursts into flames when exposed to air, and red phosphorus, a more stable form than white phosphorus
    a. Phosphorus usually forms three or five covalent bonds with other atoms
    b. The most common forms of phosphorus occur as salts of phosphoric acid
    c. Phosphorus occurs most often in nature as phosphate rocks, mostly in the form of calcium phosphate (CaH_2PO_4) and fluoroapatite ($Ca_5[PO_4]_3F$)
  2. Phosphorus is essential to human physiology
    a. Phosphorus makes up only about 1% of the total atoms in the human body; however, it comprises about 20% of the mineral matter of the body, including bones and teeth
    b. Biologically, phosphoric acid is necessary for the synthesis of genes and proteins and for the conversion of food to energy
    c. Phosphates, or salts of phosphoric acid, are essential components of nerve cells, cell membranes, enzyme systems, and physiologic buffers
  3. Phosphorus also is found in detergents, fertilizers, flame retardants, and toothpastes and as buffers in carbonated beverages to maintain a constant pH

F. **Oxygen**
  1. Oxygen is one of the two most important nonmetals in group VIA; elemental oxygen is a colorless and odorless gas that occurs in two forms—oxygen (O_2) or ozone (O_3)

a. Oxygen is the most abundant element on earth, comprising about 21% of the earth's atmosphere, 50% of the earth's crust, and 89% (by weight) of the earth's water
b. An oxygen atom has six valence electrons in the outer shell and tends to form two covalent bonds to achieve stability
c. Oxygen is an extremely strong oxidizing agent; it accepts electrons readily
d. Oxygen is highly electronegative; it forms anions easily
e. Most metals will combine with oxygen to form some type of oxide compound
2. Oxygen is essential for human physiology
a. Without oxygen, the human body usually cannot survive for more than 5 minutes; in health care, oxygen is used to treat respiratory illnesses or distress
b. Oxygen is an essential and fundamental component for many vital biological compounds, including lipids, carbohydrates, and proteins
c. Oxygen is necessary to oxidize food molecules for the production of energy
3. Oxygen in the earth's atmosphere is constantly replenished through the oxygen cycle
a. Plants take in carbon dioxide and water and combine them to produce carbohydrates, releasing oxygen to the atmosphere as a by-product
b. People and animals breathe in the oxygen, consume plant carbohydrates, and give off carbon dioxide to the air to continue the cycle
4. Oxygen is one of the most widely used elements in industry; it is used as an oxidizing agent in the conversion of iron to steel, as a oxidizing agent in sewage treatment, as a bleaching agent for paper, and as an oxidizing agent in many organic and inorganic chemical reactions

G. Sulfur
1. Sulfur is the other important nonmetal of group VIA; elemental sulfur (S_8) is an odorless, tasteless, and bright yellow solid that is insoluble in water
a. Although sulfur only makes up about 0.06% of the earth's crust, it is readily available because it occurs naturally in the elemental form
b. It also is found combined with metals as minerals, such as gypsum ($CaSO_4 \cdot 2H_2O$) and pyrite, also called fool's gold (FeS_2); sulfur also occurs in natural gas, coal, and petroleum
c. Like oxygen, sulfur tends to form two covalent bonds with other elements
d. Sulfur also has an empty orbital available in the third electron shell and can form more than two bonds, as is seen in sulfuric acid (H_2SO_4)
e. Sulfur is fairly electronegative and tends to form the sulfide anion (S^{-2})
f. Sulfur reacts with other elements similar to the way oxygen does
g. Sulfur readily forms the oxides SO_2 and SO_3, which are formed when sulfur-containing coal is burned; these oxides are major air pollutants and contributors to acid rain
2. Sulfur is important in human physiology
a. Sulfur makes up about 0.05% of the total number of atoms in the human body, but it plays an extremely important role

b. Sulfur is an essential component of human protein, acetyl coenzyme A (an enzyme necessary for the metabolism of fatty acids), and such other biological compounds as enzymes and hormones
3. Sulfur also is found in fertilizers and sulfuric acid, the world's most important industrial chemical; sulfuric acid is a very strong oxidizing agent

H. Selenium
1. The third nonmetal of group VIA, selenium is a red to black solid found in nature in minute quantities
   a. An excellent reducing agent, selenium in oxidation states $+4$ and $+6$ is also an excellent oxidizing agent
   b. Selenium forms compounds primarily by covalent bonding
2. In industry, selenium is used in photoelectric cells and as a strong oxidizing agent in the synthesis of organic substances
3. Physiologically, selenium is an essential trace element in both animals and man
   a. In cattle, sheep, and horses, dietary deficiencies of this element cause muscle disease; in sheep, such deficiencies also cause diminished reproduction
   b. Too much selenium in the diet of grazing animals produces toxic effects (such as impaired vision, muscle weakness, and respiratory failure)
   c. In man, selenium functions as part of an enzyme system that helps protect the cell membranes by preventing the oxidation of lipids
   d. Dietary deficiencies of selenium in man causes Keshan disease, a cardiac muscle disease

I. The halogens
1. The halogens (Group VIIA) include four nonmetals: fluorine, chlorine, bromine, and iodine
   a. At room temperature, fluorine and chlorine are gases; bromine is a liquid; iodine is a solid that is unique in that it sublimes at room temperature from the solid state directly to a gas
   b. As a pure element, each halogen is a diatomic molecule with the two atoms connected by a single covalent bond; however, all halogens are too reactive to occur in nature as pure elements
   c. Because of their high reactivity, the halogen elements are always found in combination with other elements
   d. Every halogen atom has seven valence electrons; halogens tend to form a single covalent bond with one other atom
   e. Halogens are fairly electronegative; in nature, halogens are almost always found as halide ions in the -1 oxidation state; the element fluorine is the most electronegative element on the periodic table
   f. Halogens are reactive oxidizing agents; because of their high electronegativity, they easily accept electrons from other atoms
   g. All halogens are toxic to humans and animals
   h. All of the halogens combine with hydrogen to form the hydrogen halides
2. Fluorine, chlorine, and iodine have important physiological functions in the human body
   a. Fluoride is an essential trace element in the structure of bones and teeth

  b. Chloride ions are part of the principal human electrolytes sodium chloride (NaCl) and potassium chloride (KCl); these electrolytes help to maintain normal internal body stability (homeostasis)
  c. Iodide ions are essential for proper thyroid gland function; thyroid hormones contain iodide and ultimately control the overall metabolic functions of the body
 3. Halogen nonmetals have various uses in industry
  a. Fluorine is used in uranium isotope separation and to make the insulating plastic Teflon
  b. Chlorine is used as a bleaching agent, as a water purification agent, and in industrial hydrochloric acid (muriatic acid)
  c. Bromine is used in gasoline additives and in photographic film
  d. Iodine is used to make medicinal antiseptics

II. Noble Gases

A. General information
1. The *noble gases* (also called *inert* or *rare* gases) compose Group 0 of the periodic table and include helium, neon, argon, krypton, xenon, and radon
2. These elements are all gases at room temperature
3. All have filled outer electron shells; they have no tendency to form bonds of any kind with another atom
4. In nature, noble gases exist as single atoms that do not combine to form molecules
5. In 1962, scientists were able to produce compounds from an inert gas; however, noble gases are still considered inert elements; they react with other atoms only in unusual circumstances

B. Uses of noble gases
1. Noble gases have no role in human physiology
2. Helium is used in the space industry to pressurize rocket fuels, as a coolant, and as a substance to dilute oxygen gas in spacecraft atmospheres
3. Neon is used in luminescent electric lighting (neon lights)
4. Argon is used in electric light bulbs

Study Activities

1. Identify all of the nonmetals on the periodic table.
2. Write the electron configurations of four different nonmetal elements.
3. Describe the biological significance of carbon, hydrogen, oxygen, and nitrogen to living organisms.
4. List and describe five nonmedical uses of halogens and noble gases.
5. Write eight balanced equations for chemical compounds involving carbon, oxygen, and hydrogen.

Appendices

Appendix A: Scientific Notation

Appendix B: Table of Elements

Appendix C: Periodic Table of Elements

Appendix D: Glossary

Selected References

Index

Appendix A: Scientific Notation

Scientific Notation is a type of shorthand for writing numbers too large or too small to use conveniently. This system expresses numbers as powers of 10. For example, one thousand is 10 x 10 x 10 or 10^3; one million is 10 x 10 x 10 x 10 x 10 x 10 or 10^6.

A number written in scientific notation consists of two parts multiplied together. The first part, called the *coefficient,* has all the significant figures written as a number between 1 and 10. The second part, called the *exponent,* is a power of 10 that returns the original value to the number. For example:

$$100 = 1 \times 10^2 \qquad 358 = 3.58 \times 10^2 \qquad 42{,}200 = 4.22 \times 10^4$$

If the exponent is a positive number, it indicates how many times the coefficient is to be multiplied by the number 10. For example,

$$7.35 \times 10^4 = 7.35 \times 10 \times 10 \times 10 \times 10 = 73{,}500$$

If the exponent is a negative number, it indicates how many times the coefficient is to be divided by the number 10. For example:

$$7.35 \times 10^{-4} = \frac{7.35}{10 \times 10 \times 10 \times 10} = 0.000735$$

See *Exponential Forms* for examples of numbers expressed in scientific notation using both positive and negative exponents.

Justification of Numbers in Scientific Notation

A *justified number* is a number in scientific notation form with a coefficient between 1 and 10. In many calculations, the final answer is an unjustified number, a number in which the coefficient is not between 1 and 10. To perform calculations smoothly, especially when adding or subtracting numbers in scientific notation, scientists must be able to move back and forth between justified and unjustified numbers easily.

The following set of numbers are all equal; however, only one number is a justified number.

$$2{,}200 \times 10^3$$
$$220 \times 10^4$$
$$22 \times 10^5$$
$$2.2 \times 10^6 \leftarrow \text{Justified number}$$
$$0.22 \times 10^7$$
$$0.022 \times 10^8$$
$$0.0022 \times 10^9$$

Calculating with Scientific Notation

To add or subtract in scientific notation, change all numbers to the same power of 10 before performing the calculation. Add or subtract the coefficients only, then justify the final answer.

For multiplication and division, altering the power of 10 is not necessary until the final answer is obtained. For multiplication, multiply the coefficients to find the new coefficient. Then, algebraically add the exponents to find the new exponent and justify the final answer. For division, divide the coefficients of the numerator by the coefficient of the denominator to find the new coefficient. Algebraically subtract the

Exponential Forms

This table shows the equivalents for numbers expressed in scientific notation using both positive and negative exponents.

NUMBER	POSITIVE EXPONENTIAL FORM	NUMBER	NEGATIVE EXPONENTIAL FORM
10	1×10^1	0.1	1×10^{-1}
100	1×10^2	0.01	1×10^{-2}
1,000	1×10^3	0.001	1×10^{-3}
10,000	1×10^4	0.0001	1×10^{-4}
100,000	1×10^5	0.00001	1×10^{-5}
23	2.3×10^1	0.23	2.3×10^{-1}
749	7.49×10^2	0.075	7.5×10^{-2}
2,280	2.28×10^3	0.00228	2.28×10^{-3}
50,200	5.02×10^4	0.000502	5.02×10^{-4}
800,000	8.0×10^4	0.000080	8.0×10^{-5}

exponent of the denominator from the exponent of the numerator; then justify the final answer.

The final answer for all calculations must contain the correct number of significant figures. Typically, the calculator answer contains far more significant figures than the initial figures justify.

Addition and subtraction

In addition and subtraction, the number of significant figures is determined by the initial number that has the lowest number of digits to the right of the decimal point.

To find the sum of $2.03 \times 10^4 + 2.6 \times 10^2 + 1.5 \times 10^5$:

Change all figures to the same power of 10.

$\ 0.203 \times 10^5$
$\ 0.0026 \times 10^5$
$+\ \underline{\ 1.5 \times 10^5\ }$
$\ 1.7056 \times 10^5$ (Calculator answer)
1.7×10^5 (Correct answer with appropriate number of significant figures)

To find the difference between 1.7×10^3 and 1.7×10^2:

Change all figures to the same power of 10.

$\ 1.7 \times 10^3$
$-\ \underline{0.17 \times 10^3}$
1.53×10^3 (Calculator answer)
1.5×10^3 (Correct answer with appropriate number of significant figures)

Multiplication and Division

In multiplication and division, the final answer has the same number of signficant figures as the initial number with the smallest number of significant figures.

To multiply (1.24×10^4) by (5.0×10^{-2}):

First multiply the coefficients.

$$1.24 \times 5.0 = 6.2 \text{ (New coefficient)}$$

To find the final exponent of 10, algebraically add the exponents of the original numbers.

$$10^4 \times 10^{-2} = 10^{(+4) + (-2)} = 10^2 \text{ (New exponent of 10)}$$

Final answer: 6.2×10^2

To divide (4.5×10^5) by (3.050×10^{-1}):

First divide the coefficient of the numerator by the coefficient of the denominator.

$$4.5 \div 3.050 = 1.4754098 \text{ (New coefficient using a calculator)}$$

To find the final exponent of 10, algebraically subtract the exponent of the denominator from the exponent of the numerator.

$$10^5 \div 10^{-1} = 10^{(+5) - (-1)} = 10^6 \text{ (New exponent of 10)}$$

Final answer with correct number of significant figures: 1.5×10^6

Appendix B: Table of Elements

This alphabetical list gives the name, symbol, atomic number, and atomic weight of all known elements. Also included is a brief explanation of each name's derivation.

NAME AND SYMBOL	ATOMIC NUMBER	ATOMIC MASS*	DERIVATION OF NAME
Actinium (Ac)	89	227.03	Greek: Aktis or Aktinos, meaning beam or ray
Aluminum (Al)†	13	26.98	Latin: Alum or Alumen, meaning astringent taste
Americium (Am)‡	95	(243)	Named for the Americas
Antimony (Sb)†	51	121.75	Latin: Stibium, meaning mark
Argon (Ar)†	18	39.95	Greek: Argos, meaning inactive
Arsenic (As)†	33	74.92	Latin: Arsenicum; Greek: Arsenikon, meaning yellow pigment
Astatine (At)	85	(210)	Greek: Astatos, meaning unstable
Barium (Ba)	56	137.34	Greek: Barys, meaning heavy
Berkelium (Bk)‡	97	(247)	Named for Berkeley, California
Beryllium (Be)†	4	9.012	Greek: Berryllos; Latin: Beryl, meaning sweet
Bismuth (Bi)	83	208.98	German: Bisemutum
Boron (B)†	5	10.81	Arabic: Buraq; Persian: Burah, meaning white
Bromine (Br)†	35	79.90	Greek: Bromos, meaning stench
Cadmium (Cd)	48	112.41	Latin: Cadmia, meaning calamine
Calcium (Ca)†	20	40.08	Latin: Calx, meaning lime
Californium (Cf)‡	98	(251)	Named for California
Carbon (C)†	6	12.01	Latin: Carbo, meaning charcoal
Cerium (Ce)	58	140.12	Named for the asteroid Ceres
Cesium (Cs)	55	132.91	Latin: Caesius, meaning sky blue
Chlorine (Cl)†	17	35.45	Greek: Chloros, meaning greenish yellow
Chromium (Cr)†	24	51.996	Greek: Chroma, meaning color
Cobalt (Co)†	27	58.93	German: Kobold, meaning goblin
Copper (Cu)†	29	63.55	Latin: Curium, from the island of Cyrpus, the ancient source of copper
Curium (Cm)‡	96	(247)	Named for Pierre and Marie Curie
Dysprosium (Dy)	66	162.50	Greek: Dyprositos, meaning hard to get at
Einsteinium (Es)‡	99	(254)	Named for Albert Einstein
Erbium (Er)	68	167.26	Named for Ytterby, a town in Sweden where many rare earths were discovered

*The atomic masses given here correspond to the 1961 values of the Commission on Atomic Weights. Masses in parentheses are those of the most stable or most common isotopes.
†These 39 elements are common.
‡These 13 elements have been made artificially; they are not found in nature. Claims have been made for the artificial production of other elements whose names are not yet established.

Table of Elements (continued)

NAME AND SYMBOL	ATOMIC NUMBER	ATOMIC MASS*	DERIVATION OF NAME
Europium (Eu)	63	151.96	Named for Europe
Fermium (Fm)‡	100	(257)	Named for Enrico Fermi
Fluorine (F)†	9	18.998	Latin and French: Fluerre, meaning flow or flux
Francium (Fr)	87	(223)	Named for France
Gadolinium (Gd)	64	157.25	Named for Johan Gadolin, a Finnish rare earth chemist
Gallium (Ga)	31	69.72	Latin: Gallia, meaning France
Germanium (Ge)	32	72.59	Latin: Germania, meaning Germany
Gold (Au)†	79	196.97	Latin: Aurum, meaning shining dawn
Hafnium (Hf)	72	178.49	Latin: Hafnia, meaning Copenhagen
Helium (He)†	2	4.003	Greek: Helios, meaning the sun
Holmium (Ho)	67	164.93	Latin: Holmia, meaning Stockholm
Hydrogen (H)†	1	1.008	Greek: Hydro, meaning water; Genes, meaning forming
Indium (In)	49	114.82	Named for indigo
Iodine (I)†	53	126.90	Greek: Iodes, meaning violet
Iridium (Ir)	77	192.22	Latin: Iris, meaning rainbow
Iron (Fe)†	26	55.85	Latin: Ferum
Krypton (Kr)	36	83.80	Greek: Kryptos, meaning hidden
Lanthanum (La)	57	138.91	Greek: Lanthanein, meaning to be hidden
Lawrencium (Lr)‡	103	(256)	Named for Ernest O. Lawrence
Lead (Pb)†	82	207.19	Latin: Plumbum, meaning heavy
Lithium (Li)†	3	6.94	Greek: Lithos, meaning stone
Lutetium (Lu)	71	174.97	Named for Lutetia, an ancient name for Paris
Magnesium (Mg)†	12	24.31	Named for Magnesia, a district in Greece
Manganese (Mn)†	25	54.94	Latin: Magnes, meaning magnet
Mendelevium (Md)‡	101	(258)	Named for Dmitri Mendeleev
Mercury (Hg)†	80	200.59	Latin: Hydrargyrum, meaning liquid silver
Molybdenum (Mo)	42	95.94	Greek: Molybdos, meaning lead
Neodymium (Nd)	60	144.24	Greek: Neos, meaning new; Didymos, meaning twin
Neon (Ne)†	10	20.18	Greek: Neos, meaning new

(continued)

Table of Elements (continued)

NAME AND SYMBOL	ATOMIC NUMBER	ATOMIC MASS*	DERIVATION OF NAME
Neptunium (Np)‡	93	237.05	Named for the planet Neptune
Nickel (Ni)†	28	58.71	German: Nickel, meaning Satan
Niobium (Nb)	41	92.91	Greek: Named for Niobe, the daughter of Tantalus
Nitrogen (N)†	7	14.01	Latin: Nitrum; Greek: Nitron, meaning native soda
Nobelium (No)‡	102	(255)	Named for Alfred Nobel
Osmium (Os)	76	190.2	Greek: Osme, meaning smell
Oxygen (O)	8	15.99	Greek: Oxys, meaning sharp or acid; Latin: Genes, meaning forming
Palladium (Pd)	46	106.42	Named for the asteroid Pallas
Phosphorus (P)†	15	30.97	Greek: Phosphoros, meaning light-bearing
Platinum (Pt)	78	195.09	Spanish: Platina, meaning silver
Plutonium (Pu)‡	94	(244)	Named for the planet Pluto
Polonium (Po)	84	(210)	Named for Poland
Potassium (K)†	19	39.10	Latin: Kalium, meaning potash
Praseodymium (Pr)	59	140.91	Greek: Prasios, meaning green; Didymos, meaning twin
Promethium (Pm)‡	61	(147)	Named for Prometheus, from Greek mythology
Protactinium (Pa)	91	231.04	Greek: Protos, meaning first; also named for the element Actinium because it disintegrates into actinium
Radium (Ra)	88	226.03	Latin: Radius, meaning ray
Radon (Rn)	86	(222)	Name derived from radium
Rhenium (Re)	75	186.21	Latin: Rhenus, meaning Rhine
Rhodium (Rh)	45	102.91	Greek: Rhodon, meaning rose
Rubidium (Rb)	37	85.47	Latin: Rubidius, meaning dark red
Ruthenium (Ru)	44	101.07	Latin: Ruthenia, meaning Russia
Samarium (Sm)	62	150.36	Named for samarskite, a mineral
Scandium (Sc)	21	44.96	Latin: Scandia, meaning Scandinavia
Selenium (Se)	34	78.96	Greek: Selene, meaning moon
Silicon (Si)†	14	28.09	Latin: Silex, meaning silicis (flint)
Silver (Ag)†	47	107.87	Latin: Argentum, meaning silver
Sodium (Na)†	11	22.99	Latin: Natrium, meaning soda
Strontium (Sr)	38	87.62	Named for Strontian, a town in Scotland

*The atomic masses given here correspond to the 1961 values of the Commission on Atomic Weights. Masses in parentheses are those of the most stable or most common isotopes.
†These 39 elements are common.
‡These 13 elements have been made artificially; they are not found in nature. Claims have been made for the artificial production of other elements whose names are not yet established.

Table of Elements (continued)

NAME AND SYMBOL	ATOMIC NUMBER	ATOMIC MASS*	DERIVATION OF NAME
Sulfur (S)†	16	32.06	Latin: Sulphurium
Tantalum (Ta)	73	180.95	Named for Tantalus, from Greek mythology
Technetium (Tc)‡	43	98.90	Greek: Technetos, meaning artificial
Tellurium (Te)	52	127.60	Latin: Tellus, meaning earth
Terbium (Tb)	65	158.93	Named for Ytterby, a town in Sweden
Thallium (Tl)	81	204.37	Greek: Thallos, meaning a budding twig
Thorium (Th)	90	232.04	Named for Thor, the Norse God of war
Thulium (Tm)	69	168.93	Named for Thule, an early name for Scandinavia
Tin (Sn)†	50	118.69	Latin: Stannum
Titanium (Ti)	22	47.90	Latin: Titans, meaning giant deities
Tungsten (W)†	74	183.85	Named for wolframite, a mineral
(Unnilennium) (Une)	109	(266)	Official name not agreed on
(Unnilhexium) (Unh)	106	(263)	Official name not agreed on
(Unniloctium) (Uno)	108	(265)	Official name not agreed on
(Unnipentium)¶ (Unp)	105	(262)	Official name not agreed on
(Unniquadium) ¶ (Unq)	104	(261)	Official name not agreed on
(Unnilseptium) (Uns)	107	(262)	Official name not agreed on
Uranium (U)†	92	238.03	Named for the planet Uranus
Vanadium (V)	23	50.94	Named for Vanadis, the Norse goddess of love
Xenon (Xe)	54	131.29	Greek: Xenon, meaning stranger
Ytterbium (Yb)	70	173.04	Named for Ytterby, a town in Sweden
Yttrium (Y)	39	88.91	Named for Ytterby, a town in Sweden
Zinc (Zn)†	30	65.38	German: Zink, meaning of obscure origin
Zirconium (Zr)	40	91.22	Arabic: Zargun, meaning gold color

¶Unnipentium is unoffically called Hahnium. Unniquadium is called Kurchatovium by Russian scientists and Rutherfordium by American scientists.

Appendix C: Periodic Table of Elements

The periodic table is an arrangement of all known elements according to atomic number and periodically related chemical properties. The vertical columns represent groups of elements with similar chemical properties. The horizontal columns represent the period-

Groups →

Periods	1(IA)	2(IIA)	3(IIIB)	4(IVB)	5(VB)	6(VIB)	7(VIIB)	8(VIII)	9(VIII)
one	1 H 1.00797								
two	3 Li 6.941	4 Be 9.01218							
three	11 Na 22.98977	12 Mg 24.305							
four	19 K 39.098	20 Ca 40.08	21 Sc 44.9559	22 Ti 47.90	23 V 50.9414	24 Cr 51.996	25 Mn 54.9380	26 Fe 55.847	27 Co 58.9332
five	37 Rb 85.4678	38 Sr 87.62	39 Y 88.9059	40 Zr 91.22	41 Nb 92.9064	42 Mo 95.94	43 Tc 98.9062	44 Ru 101.07	45 Rh 102.905
six	55 Cs 132.9054	56 Ba 137.34	57–71	72 Hf 178.49	73 Ta 180.9479	74 W 183.85	75 Re 186.2	76 Os 190.2	77 Ir 192.22
seven	87 Fr (223)	88 Ra 226.0254	89-103	104 Unq (261)	105 Unp (262)	106 Unh (263)	107 Uns (262)	108 Uno (265)	109 Une (266)

Example: 2 He 4.00260 — atomic number, chemical symbol, atomic mass

Transition Elements

Inner-transition Elements

Lanthanides (rare earths)

| 57 La 138.9055 | 58 Ce 140.12 | 59 Pr 140.9077 | 60 Nd 144.24 | 61 Pm (147) | 62 Sm 150.36 | 63 Eu 151.9 |

Actinides

| 89 Ac 227.0278 | 90 Th 232.0381 | 91 Pa 231.0359 | 92 U 238.0289 | 93 Np 237.0482 | 94 Pu (244) | 95 Am (243) |

icity, or repeating nature, of the elements and contain elements of relatively similar weight. The point of separation between metallic and nonmetallic elements is approximate.

Groups ───────────────────────────────→ Noble Gases

			13(IIA)	14(IVA) Nonmetals 15(VA)	16(VIA)	Halogens 17(VIIA)	18(0) 2 **He** 4.00260
			5 **B** 10.81	6 **C** 12.01115	7 **N** 14.0067	8 **O** 15.9994	9 **F** 18.99840	10 **Ne** 20.179
Transition Elements 10(VIII)	11(IB)	Metals 12(IIB)	13 **Al** 26.98154	14 **Si** 28.086	15 **P** 30.97376	16 **S** 32.06	17 **Cl** 35.453	18 **Ar** 39.948
28 **Ni** 58.71	29 **Cu** 63.546	30 **Zn** 65.38	31 **Ga** 69.72	32 **Ge** 72.59	33 **As** 74.9216	34 **Se** 78.96	35 **Br** 79.904	36 **Kr** 83.80
46 **Pd** 106.4	47 **Ag** 107.868	48 **Cd** 112.41	49 **In** 114.82	50 **Sn** 118.69	51 **Sb** 121.75	52 **Te** 127.60	53 **I** 126.9045	54 **Xe** 131.30
78 **Pt** 195.09	79 **Au** 196.9665	80 **Hg** 200.59	81 **Tl** 204.37	82 **Pb** 207.19	83 **Bi** 208.9804	84 **Po** (210)	85 **At** (210)	86 **Rn** (222)

Inner-transition Elements

64 **Gd** 157.25	65 **Tb** 158.9254	66 **Dy** 162.50	67 **Ho** 164.9304	68 **Er** 167.26	69 **Tm** 168.9342	70 **Yb** 173.04	71 **Lu** 174.967
96 **Cm** (247)	97 **Bk** (247)	98 **Cf** (251)	99 **Es** (254)	100 **Fm** (257)	101 **Md** (258)	102 **No** (255)	103 **Lr** (256)

Appendix D: Glossary

Absolute zero — temperature on the Kelvin scale below which no substance can be cooled (0° K)

Acid — substance that ionizes in water to produce H^+ ions; a proton donator

Activation energy — minimum amount of kinetic energy with which two particles must collide for a chemical reaction to occur

Alkali metals — metals found in Group I of the periodic table

Alkaline earth metals — metals found in Group II of the periodic table

Alpha radiation — radiation particles with a charge of +2 and a mass of 4

Amorphous solid — solid composed of a fixed, randomly repeating arrangement of atoms or molecules

Anion — ion with a negative charge; an ion attracted to the positive pole (anode) of an electrical circuit

Atom — smallest particle of an element that can exist and still have the properties of the element

Atomic mass — weighted average of all the different isotopes of an element; a reflection of the total number of protons and neutrons in the nucleus of an element; also called *atomic weight*

Atomic mass unit — unit for measuring the total mass of neutrons and protons in a nucleus based on the assignment of exactly 12 atomic mass units (amu) to the mass of the carbon-12 isotope

Atomic number — number of protons in the nucleus of an atom; each element has a unique, characteristic atomic number

Atomic symbol — letter or letters that represent an element, especially when elements are combined in compounds

Atomic weight — weighted average of the masses of the naturally occurring isotopes of an element, as compared to carbon and expressed in atomic mass units; also called *atomic mass*

Aufbau principle — rule of chemistry that states as protons are added one by one to the nucleus to form the various elements, electrons similarly are added to the atomic orbitals

Avogadro's law — principle of gas behavior that states equal volumes of gases at the same temperature and pressure contain equal numbers of molecules

Avogadro's number — number of atoms in the carbon 12 isotope (^{12}C), equal to 6.02×10^{23} atoms; the standard number for determining the equivalent of one mole

Base — substance that ionizes in water to produce OH^- ions; a proton acceptor

Beta radiation — radiation particles with a charge of -1 and a mass of essentially 0

Boiling point — temperature at which a substance changes from a liquid to a gas

Boyle's law — principle of gas behavior that states at a constant temperature the volume of a gas varies inversely with its pressure; $P_1V_1 = P_2V_2$

Buffer — solution that resists sudden changes in pH when acids or bases are added to it; usually a solution of a weak acid and its conjugate base or a weak base and its conjugate acid

Calorie — amount of heat energy required to raise the temperature of one gram of water one degree Celsius, from 14.5° C to 15.5° C

Catalyst — substance that reduces the activation energy required to initiate and sustain a reaction; a substance that increases the rate of a chemical reaction without being consumed in the reaction

Cation — ion with a positive charge; an ion attracted to the negative pole (cathode) of an electrical circuit

Charles' law — principle of gas behavior that states at a constant pressure, the volume of a gas varies directly with its temperature; $V_1/T_1 = V_2/T_2$

Chelate — complex molecule in which a metal ion or atom is bound at two or more points to another molecule, forming a ringlike structure

Chemical bond — force that holds atoms or ions together to form a compound

Chemical change — change in the basic chemical composition of matter to form a different substance

Chemical equilibrium — dynamic state in which the forward rate of a chemical reaction is the same as the reverse rate of the reaction

Chemical formula — symbolic representation of the atoms or elements comprising a compound

Chemical properties — properties exhibited by matter when it changes its chemical composition

Chemical reaction — reaction involving atoms or molecules in which a new compound is formed

Colligative properties — physical properties that vary with the number of particles present in a solution rather than with the solution's chemical nature; for example, the boiling point of a solution is a colligative property

Colloid — heterogeneous mixture that usually, but not always, has a liquid solvent and a solute particle size significantly greater than that of a solution

Compound — pure substance in which two or more elements are united in specific proportions

Compressibility — measure of the decreased volume of a substance when pressure is applied to it

Conjugate acid — acid formed when a base accepts a proton from another molecule

Conjugate base — base formed when an acid gives up a proton to another molecule

Cosmic radiation — subnuclear particles originating from the sun and outer space that continually bombard the earth

Covalent bond — chemical bond formed when two atoms share electrons in mutually held molecular orbitals

Crystalline lattice — repetitious three-dimensional arrangement of atoms found in a crystalline solid

Crystalline solid — solid consisting of particles arranged in a regularly repeating, three-dimensional pattern

Dalton's law of partial pressures — principle of gas behavior that states the total pressure of a gas mixture is equal to the sum of the partial pressures of all the component gases; $P_{total} = P_1 + P_2 + P_3 + P_4 + P_5 + \ldots$

Density — mass of an object divided by its volume

Diffusion — spontaneous mixing of particles from an area of higher concentration to one of lower concentration; the movement of components of a gaseous system as a result of differences in concentration

Electromagnetic energy — energy that travels through space in regularly spaced waves, much as water travels across the ocean in waves

Electron — negatively charged particle that orbits around the nucleus of an atom and that has a relative mass of $1/1,837$ amu.

Electronegativity — ability of an atom to attract electrons toward itself in a chemical bond

Electrostatic attraction — attractive force between two oppositely charged ions

Element — pure substance that cannot be decomposed into simpler substances by ordinary chemical means

Endothermic reaction — reaction in which energy in the form of heat is required to initiate the reaction and keep it going

Energy — measure of the capacity to do work

Energy level — arbitrary portion of the electromagnetic spectrum and the energy contained within it (for example, the visible light portion of the electromagnetic spectrum includes a specific range of energy levels); one of several orbits or shells in which electrons travel around the nucleus of an atom

English system — measurement system that uses various bases for all measurements; also called the *U.S. Customary system*

Equivalent — specific quantity of reactant needed to supply one mole of reacting units of a chemical reaction (for acids, an equivalent is the amount of acid that will supply one mole of H^+ ions; for bases, the amount of base that will accept one mole of H^+ ions; for oxidizing agents, the amount of substance that will accept one mole of electrons in an oxidation-reduction reaction; for reducing agents, the amount of substance that will provide one mole of electrons in an oxidation-reduction reaction)

Evaporation — process by which molecules of a liquid escape from the surface of the liquid and become a gas

Exothermic reaction — reaction in which heat is given off to the surroundings as the products of the reaction are formed

Expansivity — measure of the extent a substance will increase in volume as a result of a decrease in applied pressure

Formula weight — sum of the atomic weights of all the atoms in a molecule

Freezing point — temperature at which a substance changes from a liquid to solid state; synonymous to melting point

Gamma radiation — high-energy electromagnetic radiation similar to X-rays

Gamma rays — high-energy radiation with no charge and no mass emitted from the nuclei of unstable atoms

Gas — physical state of matter that has no shape, diffuses readily, and assumes the full volume of any closed container

Gay-Lussac's law — principle of gas behavior that states at constant volume, the pressure of a gas varies directly with its temperature; $P_1/T_1 = P_2/T_2$

General gas law — algebraic combination of Boyle's Law, Charles' Law, and Gay-Lussac's Law; $P_1V_1/T_1 = P_2V_2/T_2$

Graham's law of effusion — principle of gas behavior that states the effusion rate of a gas is inversely proportional to the square root of its molecular weight; $\frac{r_1}{r_2} = \sqrt{\frac{M_2}{M_1}}$

Gram equivalent — for acids, the weight of acid in grams that will supply one mole of H^+ ions; for bases, the weight of base in grams that will supply one mole of OH^- ions; for oxidizing or reducing agents, the weight of compound in grams that will supply or accept one mole of electrons

Ground state — lowest energy level an electron normally inhabits in a given atom

Half-life — time required for one-half of a given amount of a radionuclide to decay

Halogen — any of the elements in group VII of the periodic table

Heat energy — energy that is transferred from one place to another because of a difference in temperature

Heat of fusion — amount of heat energy required to change one gram of a substance from a solid to a liquid at its freezing (or melting) point

Heat of vaporization — amount of heat energy required to change one gram of a substance from a liquid to a gas at its boiling point

Henry's law — principle of gas behavior that mathematically states the solubility of a gas in a liquid is directly proportional to the pressure of the gas over the liquid

Heterogeneous mixture — mixture in which component substances are in visibly different parts or phases with nonuniform characteristics throughout the mixture

Homogeneous mixture — mixture in which all component substances are in the same visible part or phase with uniform characteristics throughout the mixture

Hydrocarbon — organic compound containing only carbon and hydrogen atoms

Hydrogen bonding — special type of intermolecular bonding that occurs when the hydrogen in a molecule forms additional loose bonds with electronegative atoms in neighboring molecules

Hydrolysis — any reaction in which water is one of the reactants; more specifically, the combination of water with a salt to produce an acid and a base

Ideal gas law — algebraic combination of Boyle's law, Charles' law, and Avogadro's law; $PV = nRT$

Inert gases — see *Noble gases*

Inner-transition elements — elements with atomic numbers 58 to 71 and 90 to 103, located in the two bottom rows of the periodic table

Ion — atom that assumes an electrical charge as a result of having gained or lost one or more electrons

Ionic bond — chemical bond formed as a result of attraction between two oppositely charged ions

Ionic compound — compound consisting of an orderly arrangement of oppositely charged ions that are combined in a ratio such that the compound is electrically neutral

Ionic solid — solid whose atoms are bound together by the strong forces of attraction between oppositely charged ions

Ionization constant — number indicating the degree to which an acid or base dissociates in water at equilibrium; K_a or K_b

Ionization energy — energy required to remove an electron from an atom

Isotope — one of several different forms of an atom, each of which has a characteristic mass number caused by a variation in the number of neutrons in the atom's nucleus; also called a *nuclide*

Kinetic energy — energy that an object or particle possesses as a result of its mass and motion

Law of conservation of mass — principle of chemistry and physics that states that matter is neither created nor destroyed in an ordinary chemical reaction

Law of conservation of matter and energy — principle that states matter may be converted to energy but the total amount of matter and energy in the universe remains the same

Liquid — physical state of matter that assumes the shape of its container; a compound or element with no defined shape and whose molecules or atoms are relatively close together but have much freedom of movement

Mass — measure of the quantity of matter in an object

Mass number — total number of protons and neutrons in the nucleus of the atom

Matter — anything that occupies space and has mass

Melting point — temperature at which a substance in its solid state changes into a liquid state; also called *freezing point*

Metal — group of elements composing nearly three-fourths of the elements on the periodic table and having certain common physical characteristics, including malleability, ductility, a shiny surface, and a high melting point

Metallic bond — unique bond formed by metals in which free negatively charged electrons are attracted to positively charged atomic centers

Metalloid — group of elements that have some of the characteristics of a metal, such as the ability to conduct electricity, but that lack other metal characteristics, such as a shiny surface

Metric system — most commonly used measurement system in science in which units are based on multiples of ten

Mixture — physical combination of two or more substances (either elements or compounds, or both) each of which retains its original identity; mixtures may be separated into their component substances through physical means

Molality — number of moles of solute in a kilogram of solvent

Molarity — in a solution, the number of moles of solute in a liter of solution

Molar weight — weight in grams of one mole of atoms or molecules (for atoms, the molar weight is the atomic weight in grams; for molecules, the formula weight in grams)

Mole — unit of measure that is equal to 6.02×10^{23} particles of a substance, usually expressed as grams/mole; unit equivalent to the atomic weight or formula weight of a substance

Molecular solid — solid whose molecules are held together by relatively weak intermolecular forces, such as van der Waals forces or hydrogen bonding

Molecule — two or more atoms tightly bound together and functioning as a single unit

Neutralization reaction — reaction of an acid with a base to form a salt plus water

Neutron — nuclear particle with no charge and one atomic mass unit (amu)

Noble gases — all of the elements of Group 0; also called the *rare gases* or *inert gases;* the only elements on the periodic table that occur as monatomic gases

Nonmetals — certain elements from Groups IVA, VA, and VIA, plus hydrogen and all of the elements in Group VIIA on the periodic table

Normality — number of equivalents of solute in one liter of solution

Nuclear decay — process in which an unstable atomic nucleus gives off radioactive particles or waves to become a stable nucleus

Nuclear fission — splitting of an atomic nucleus into two or more lighter nuclei by bombardment with incoming high-speed particles, such as neutrons

Nuclear fusion — combining of two small nuclei to form a larger, more stable nucleus

Nuclear transmutation — any process that results in the creation of a new nucleus, either by radioactive decay or by bombarding an existing element with neutrons, electrons, or other nuclei

Nuclide — see *Isotope*

Octet rule — tendency of elements in Groups I to VII of the periodic table to form bonds that result in eight valence electrons in the outer energy level of each atom

Osmosis — movement of water through a semipermeable membrane from a region of low solute concentration to a region of high solute concentration

Oxidation — loss of electrons by an atom

Oxidation number — total charge an atom would have in a molecule if electrons were transferred completely to the atoms with the highest electronegativity

Oxidation-reduction reaction — reaction in which electrons are transferred from one reactant to another

Oxidizing agent — substance that accepts electrons in an oxidation-reduction reaction; a substance that causes the oxidation of a reactant molecule

Partial pressure — pressure that a gas in a gas mixture would exert if it were present alone in the same conditions

Percent solution — concentration of solute in a given solvent expressed in terms of volume/volume, weight/weight, or weight/volume

Periodic law — principle of chemistry that states when the elements are arranged according to increasing atomic number, their chemical and physical properties show periodic similarities

Periodic table — arrangement of the elements according to periodicity or the repeating nature of the chemical properties of the elements

pH system — numeric system of designating the concentration of H_3O^+ ions in an aqueous solution; mathematically, $pH = -\log [H^+]$

Physical change — change in matter from one state to another, as from a liquid to a gas or solid

Physical properties — properties of matter, such as melting point, color, or odor, that can be observed or measured without changing the substance into another substance

Polarity — measure of inequality in the sharing of bonding electrons; the existence of partial electrical charge in two different regions of a molecule as a result of uneven distribution of bonding electrons

Polyatomic ion — group of atoms held together by strong covalent bonds and functioning as a single unit

Potential energy — stored energy associated with the attraction or repulsion of objects or particles

Precipitation — formation of a solid in solution

Product — any final substance (or substances) resulting from a chemical reaction

Proton — positively charged nuclear particle with a relative mass of 1 atomic mass unit (amu)

Pure substance — form of matter with a definite, constant, and unique composition and distinct properties

Quantum — discrete quantity of energy that is emitted or absorbed when an electron moves from one energy level to another

Rad — unit of radiation dosage used to describe the amount of energy absorbed by irradiated tissue

Radioactivity — spontaneous emission of radiation by unstable atomic nuclei, resulting in the formation of a new element

Radioisotope — isotope possessing an unstable nucleus that spontaneously emits energy in the form of radiation; also called *radionuclide*

Reactant – any starting substance that combines with another substance in a chemical reaction to produce a product

Reaction rate – speed at which a chemical reaction occurs under given circumstances, such as at a certain temperature or under particular light conditions

Reducing agent – substance that donates electrons in an oxidation-reduction reaction; a substance that causes the reduction of a reactant molecule

Reduction – gain of electrons by an atom

Rem – unit of absorbed dose of radiation that will produce the same biological effect as 1 rad of therapeutic X-rays

Representative elements – elements occupying groups I through VII and group 0 of the periodic table

Salt – ionic compound formed by the reaction between an acid and a base

Scientific notation – mathematical system in which numbers are expressed as a coefficient multiplied by powers of 10; used for ease of computation with extremely large or small numbers

Shell – energy level or orbit in which electrons travel around the nucleus of an atom

Significant figure – number of meaningful digits in a measured or calculated quantity; all the numbers of a measurement that can be read with confidence plus one last figure that is an estimate

Solid – physical state of matter that holds its shape and volume even when not in a container; a compound whose molecules are closely packed together and have little freedom of movement

Solubility – maximum amount of a substance that can be dissolved in a given quantity of water at a specific temperature; the degree to which a given solute will dissolve in a given solvent with respect to that solvent

Solute – component of a solution that constitutes the smallest part of the mixture

Solution – homogeneous mixture of two or more substances; commonly, a homogeneous mixture of a solid in a liquid, but also can be a liquid in a liquid, a gas in a liquid, or a gas in a gas

Solvent – component of a solution that constitutes the largest part of the mixture

Specific gravity – ratio of the mass of a substance compared to the mass of an equal volume of water

Specific heat – amount of heat required to raise the temperature of one gram of liquid one degree Celsius

Sublimation – change of state in which a substance passes from the solid state directly to the gaseous state at the same temperature

Surface tension – measure of the resistance of molecules at the surface of a liquid to expansion of the liquid upward and outward

Surfactant – chemical that artifically decreases the surface tension of a liquid; also called a *wetting agent*

Suspension – heterogeneous mixture in which an insoluble solid is dispersed in a liquid; solute particles in suspensions are larger than those in colloids

Systeme Internationale – modernized version of the metric system using standardized units for all types of measurement; the preferred system of measurement for health care

Temperature – measure of the average kinetic energy of the particles of a substance

Titration – laboratory procedure for measuring the unknown concentration of an acidic or basic solution

Trace metal – element required in minute amounts for normal cell growth and development; also called a *trace element*

Transition element – any element located between Groups II and III in Periods 4, 5, and 6 of the periodic table

Transition metal – any metal from Group IIIB through Group IIB on the periodic table

Triple point of water – temperature on the Kelvin scale at which liquid water, ice, and water vapor exist together in the absence of air or another substance (273.16° K)

Valence electron – one of eight electrons in the outermost electron shell of an atom; the electrons that determine the types of bonds an atom will form

van der Waals forces – intermolecular forces of attraction resulting from the temporary dispersion of the electrons in a molecule toward one end of the molecule

Vapor pressure – measure of a the tendency of a liquid to evaporate; measurable pressure exerted by the molecules of a liquid that have escaped the surface of the liquid to become a gas

Viscosity – measure of the ease of liquid flow; the higher the viscosity, the less easily the liquid flows

Weight – indication of how strongly the earth's gravity pulls on a given mass

Work – energy released or absorbed through mechanical means

X-ray diffraction – characteristic scattering of X-rays by the units of a regular crystalline solid that enables identification of the solid

Selected References

Bloomfield, M.M. *Chemistry and the Living Organism,* 4th ed. New York: John Wiley & Sons, Inc., 1987.

Bloomfield, M.M. *Study Guide for Chemistry and the Living Organism,* 4th ed. New York: John Wiley & Sons, Inc., 1987.

Brescia, F., et al. *General Chemistry,* 5th ed. New York: Harcourt Brace Jovanovich, Inc., 1988.

Campbell, J.M., and Campbell, J.B. *Laboratory Mathematics: Medical and Biological Applications,* 4th ed. St. Louis: Mosby-Year Book, Inc. 1990.

Chang, R. *Chemistry,* 3rd ed. New York: McGraw-Hill Publishing Co., 1988.

Herron, J.D. *Understanding Chemistry,* 2nd ed. New York: Random House, 1986.

Lide, D.R., ed. *CRC Handbook of Chemistry and Physics,* 71st ed. CRC Press, 1990.

Malone, L.J. *Basic Concepts of Chemistry,* 3rd ed. New York: John Wiley & Sons, Inc., 1989.

Schugar, G., and Ballinger, J. *Chemical Technicians' Ready Reference Handbook,* 3rd ed. New York: McGraw-Hill Book Co., 1990.

Tietz, N.W., ed. *Textbook of Clinical Chemistry.* Philadelphia: W.B. Saunders Co., 1986.

Watkins, K.W. *Study Guide for Chang's Chemistry,* 3rd ed. New York: Random House, 1988.

Index

A
Absolute zero, 8, 112
Acid-base indicators, 85
Acidosis, 87
Acids, 79, 81t, 112
  binary, 80
  conjugate, 80
  ionization of, 80-81
  polyprotic, 80
  reactions with bases, 81-83
  strong, 80
  ternary, 80
  theories explaining
    behavior of, 79-80
  weak, 80
  See also Bases; Buffers; pH;
    and Titration.
Actinide series, 24
Activation energy, 35, 112
Adhesive force, 53
Alkali metals, 24, 89-90, 112
Alkaline earth metals, 24, 90, 112
Alkalosis, 87
Allotropy, 49
Alpha radiation, 72, 75t, 112
Amorphous solid, 47-48, 112
Anion, 25, 112
Antimony, 94
Aqueous (electrolyte) solution, 58
Arrhenius theory, 79-80
Arsenic, 93-94
Astatine, 94
Atmosphere, 63
Atom, 2, 9, 16-17, 112
Atomic mass, 18, 112
Atomic mass units, 17
Atomic number, 17, 112
Atomic symbol, 16, 112
Atomic weight. *See* Atomic mass.
Aufbau principle, 21, 22t, 112
Avogadro's law, 66, 112
Avogadro's number, 42, 112

B
Barometer, 63
Bases, 79
  conjugate, 80
  ionization of, 80-81
  reactions with acids, 81-83
  theories explaining behavior of, 79-80
  See also Acids; Buffers; pH;
    and Titration.
Becquerel, 73
Beta radiation, 72, 75t, 112
Biochemistry, 1
Boiling point, 12, 54, 112
Boiling point elevation, 58
Bond. *See* Chemical bond.
Boron, 93
Boyle's law, 64, 64t, 112
Brønsted-Lowry theory, 79-80
Brownian movement, 59
Buffers, 87, 112

C
Calorie, 7, 13, 112
Carbon, 97
Catalyst, 36, 112
Cation, 25, 112
Celsius scale, 8
Chain reaction, 75
Charles' law, 65, 65t, 112
Chelate, 91, 92i, 112
Chemical bond, 26, 112
  covalent, 29-30, 113
  ionic, 27-29, 114
  metallic, 30-31, 114
  octet rule for, 26-27, 115
Chemical change, 11, 112
Chemical equation, 35, 38
  balancing of, 38-40
  symbols for, 39t
Chemical equilibrium, 40, 112
Chemical formula, 31, 112
  comparison of, 33t
  condensed structural, 33
  empirical, 33
  formula weight, 34
  molecular, 33
  structural, 33
Chemical properties, 12, 112
Chemical reaction, 35, 112
  endothermic, 36
  exothermic, 36
  oxidation-reduction, 36-38, 37t, 115
  reaction rate, 35-36
Chemistry, 1
Coefficient, 39
Cohesive force, 53
Colligative properties, 57, 112
Colloids, 56, 59, 59t, 113
Compound, 2, 9-10, 26, 31, 113
  covalent, 29, 30
  formula weight of, 34
  ionic, 27
  metallic, 30, 31
  rules for naming, 31-33, 32t
Compressibility, 53, 61, 113
Condensation, 54
Conjugate acid, 80, 113
Conjugate base, 80, 81t, 113
Cosmic radiation, 72-73, 113
Covalent bond, 29, 113
  nonpolar, 30
  polar, 30
Covalent compound, 29
Covalent crystals, 51
Covalent molecular compound, 29
Crystalline lattice, 27, 28i, 113
Crystalline solid, 47, 113
Crystals, 48-49
  covalent, 51
  ionic, 50
  metallic, 51
  molecular, 51
  physical properties of, 49t
  structure of, 49
  unit cells, 49, 50i
Curie, 73

D
Dalton's law of partial pressures, 68, 68t, 113
Daughter nucleus, 71-72
Density, 7, 12, 113
  of gases, 63-64
  of liquids, 53
Diffusion, 61-62, 113
Dilution, 45, 46t
Distillation, 58
Dynes, 7

E
Effusion, 68-69, 69t
Electrolyte, 58
Electrolyte (aqueous) solution, 58
Electromagnetic energy, 13-15, 113
Electromagnetic spectrum, 14i, 14-15
Electron, 17, 19, 113
  Bohr's model of, 19-20
  configurations of, 20-21, 21t
  energy levels, 19-20
  valence, 24, 26
Electronegativity, 22-23, 28-29, 113
Electrostatic attraction, 27, 113
Elements, 2, 9, 16, 113
  classification by properties, 24-25
  components of, 16-17
  identifying, 17-18
  inner-transition, 24
  representative, 24, 116
  table of, 106-109
  transition, 24, 116
Endothermic reaction, 36, 113
Energy, 7, 12, 113
  electromagnetic, 13t, 13-15, 113
  heat, 13
  kinetic, 12-13, 35, 114
  law of conservation of, 15
  potential, 13, 115
Energy levels (shells), 19-21, 113
English system, 6, 113
Equilibrium, 40
Equivalence point, 44-45, 86
Equivalent, 44-45, 86, 113
Evaporation, 54, 113
Exothermic reaction, 36, 113
Expansivity, 61, 113

F
Fahrenheit scale, 8
Fission (nuclear), 75
Force, 6-7
Formula. *See* Chemical formula.
Formula weight, 34, 113
Fractional distillation, 58
Freezing (melting) point, 12, 54, 113
Freezing point depression, 57
Frequency, 14
Fusion (nuclear), 75-76

i refers to an illustration; t, to a table.

Index

G
Gamma radiation, 72, 75t, 113
Gases, 10, 61-62, 113
  Avogadro's law, 66, 112
  Boyle's law, 64-65, 112
  Charles' and Gay-Lussac's laws, 65, 65t, 112, 113
  Dalton's law of partial pressures, 68, 68t, 113
  density of, 63-64, 113
  general gas law, 66, 66t, 113
  Graham's law of effusion, 68-69, 69t, 113
  Henry's law, 69, 70t, 114
  ideal gas law, 67, 67t, 114
  kinetic molecular theory of, 62
  pressure in, 62-63
Gay-Lussac's law, 65, 113
General gas law, 66, 66t, 113
Germanium, 93
Graham's law of effusion, 69, 69t, 113
Gram equivalent, 45, 86, 113
Ground state, 25, 113

H
Half-life, 73-74, 114
Halogens, 24, 100-101, 114
Hardness, 12
Heat energy, 13, 114
Heat of fusion, 54, 114
Heat of vaporization, 54, 114
Henry's law, 57, 69, 70t, 114
Hydrocarbons, 97, 114
Hydrogen, 25, 96-97
Hydrogen bonding, 48, 114
Hydrolysis (salt), 83, 114

I
Ideal gas law, 67, 67t, 114
Immiscible liquids, 57
Inert gases. *See* Noble gases.
Infrared light, 14
Inorganic chemistry, 1
Insoluble liquid, 57
Intermolecular forces, 48
Ion, 25, 114
Ionic bond, 27, 114
Ionic compound, 27, 27-29, 114
Ionic crystals, 50
Ionic solid, 48, 114
Ionization constant, 81, 83t, 114
Ionization energy, 22-23, 114
Ionizing radiation, 76
Isotopes (nuclides), 17-18, 18t, 71-72, 114

J, K
Joule, 7
Kelvin scale, 8, 65
Kilocalorie, 13
Kinetic energy, 12-13, 35, 114

L
Lanthanide series, 24
Lattice point, 48

Law of conservation of mass, 38, 114
Law of conservation of matter and energy, 15, 114
Liquid mixtures, 56, 56t
  colloids, 58-59, 59t, 113
  liquid solutions, 56-58
  suspensions, 59, 116
Liquids, 10, 114
  boiling point of, 54
  compressibility of, 53
  density of, 53
  freezing point of, 54
  surface tension in, 53
  viscosity of, 52-53
  See also Liquid mixtures *and* Water.
Liquid solutions, 56
  dilute, 57
  electrolyte, 58
  fractional distillation of, 58
  osmotic pressure in, 58
  physical properties of, 57
  saturated, 57
  solubility of, 56-57
  solute addition to, 57-58
  supersaturated, 57
Liter, 6

M
Manometer, 63
Mass, 6, 114
Mass number, 17
Matter, 1-2, 9-10, 114
  law of conservation of, 15
  properties of, 11-12
  states of, 10-11, 11t
Measurement, 2
  accuracy of, 2
  density, 7
  energy, 7
  force, 6-7
  precision of, 2
  pressure, 7, 63
  scientific notation in, 4-5, 103-105, 104t
  significant figures in, 3-4
  specific gravity, 7
  temperature, 8
  units of, 5-6
  velocity, 6
  volume, 6
Melting (freezing) point, 12, 54
Metallic bond, 30-31, 114
Metallic crystals, 51
Metalloids, 25, 92, 114
  antimony, 94
  arsenic, 93-94
  astatine, 94
  boron, 93
  germanium, 93
  polonium, 94
  silicon, 93
  tellurium, 94
Metals, 24, 114
  alkali, 89-90, 112
  alkaline earth, 90, 112
  classifying, 88-89, 89t
  Group IIIA, 91-92

Metals *(continued)*
  Group IVA, 92
  physical characteristics of, 88
  transition, 90-91, 116
Metric system, 5, 5t, 114
  conversion table, 6t
Microwaves, 14
Miscible liquids, 57
Mixture, 10, 114
  heterogeneous, 10, 56
  homogeneous, 10, 56
  See also Liquid mixtures.
Molality, 44, 114
Molarity, 44, 114
Molar weight, 42-43, 115
Mole, 42-43, 115
Molecular crystals, 51
Molecular solid, 48, 115
Molecule, 2, 9, 26, 115

N
Neutralization reaction, 83, 115
Neutron, 17, 115
Newtons, 7
Nitrogen, 97-98
Nitrogen fixation, 98
Noble (inert) gases, 24, 61, 101, 115
Nonelectrolyte, 58
Nonmetals, 24, 115
  carbon, 97
  halogens, 100-101
  hydrogen, 96-97
  nitrogen, 97-98
  oxygen, 98-99
  phosphorus, 98
  physical and chemical properties of, 96
  selenium, 100
  sulfur, 99-100
Normality, 44-45, 86, 115
Nuclear reactions
  balancing equations, 75
  decay, 74, 115
  fission, 75, 115
  fusion, 75-76, 115
  transmutation, 76, 115
Nuclides (isotopes), 17-18, 71-72, 115

O
Octet rule, 26-27, 115
Orbitals, 21
Organic chemistry, 1
Osmosis, 58, 115
Osmotic pressure, 58
Oxidation, 36, 115
Oxidation numbers, 37t, 37-38, 115
Oxidation-reduction reaction, 36-38, 40, 115
Oxidizing agent, 36, 115
Oxygen, 98-99

P
Pascal, 7
Percent ionization, 81, 82t

---

i refers to an illustration; t, to a table.

Index

Percent solution, 43-44, 115
Periodic law, 22, 115
Periodic table, 21, 23t, 110-111i, 115
  element classification by properties, 24-25
  groups, 23-24
  inner-transition elements, 24
  metalloids, 92-94
  metals, 88-92
  noble gases, 101
  nonmetals, 96-101
  periodic law, 22-23
  periods, 23
  transition elements, 24
pH meter, 85
pH system, 84, 115
  of common fluids, 86t
  scale, 84-85, 85t
Phosphorus, 98
Physical change, 10, 115
  endothermic, 10
  exothermic, 11
  sublimation, 11
Physical chemistry, 1
Physical properties, 11-12, 115
  of gases, 61-64
  of liquids, 52-54, 57
  of metalloids, 92-93
  of metals, 88
  of nonmetals, 96
Polarity, 54, 115
Polonium, 94
Polyatomic ions, 32, 32t, 115
Potential energy, 13, 115
Precipitation, 59, 115
Pressure, 7
  gas, 62-63
Products, 38, 115
Proton, 17, 115
Pure substance, 9, 115

Q, R

Quantum, 20, 115
Rad, 77, 115
Radiation, 72
  absorption by tissues, 77
  alpha, 72, 75t, 112
  beta, 72, 75t, 112
  cosmic, 72-73, 113
  gamma, 72, 75t, 113
  in health care, 77-78
  interaction with matter, 76-77
  ionizing, 76
  recording mass and charges of, 75, 75t
  units of, 73
Radioactivity, 71-72, 115
  alpha radiation, 72, 75t, 112
  beta radiation, 72, 75t, 112
  cosmic radiation, 72-73, 113
  gamma radiation, 72, 75t, 113
  half-life, 73-74
  units of radiation, 73
  See also Nuclear reactions *and* Radiation.
Radioisotopes (radionuclides), 71, 73t, 115
  half-life of, 73-74

Radiowaves, 14
Reactant, 35-36, 116
Reaction rate, 35-36, 116
Redox equation, 36-38, 40
Reducing agent, 37, 116
Reduction, 36, 116
Relative mass, 18
Rem, 77, 116
Representative elements, 24, 116
Roentgen, 76-77

S

Salt, 81, 82, 116
  acid salts, 83
  hydrolysis, 83
  neutralization reaction, 83, 115
  oxoacid salts, 83
Scientific notation, 2, 4-5, 103, 116
  calculating with, 103-105
  exponential forms, 104t
  justification of numbers in, 103
Selenium, 100
Shells. *See* Energy levels.
Significant figures, 3t, 3-4, 4t, 115
Silicon, 93
Solids, 10, 47, 116
  amorphous, 47-48
  crystalline, 47
  ionic, 48
  molecular, 48
  See also Crystals.
Solubility, 12, 56-57, 116
Solute, 43, 116
Solution, 43, 116
  dilution of, 45, 46t
  molality of, 44
  molarity of, 44
  normality of, 44-45
  percent solution, 43-44
Solvent, 43, 116
Specific gravity, 7, 116
Specific heat, 55, 116
Sublimation, 11, 116
Subshells, 20
Sulfur, 99-100
Surface tension, 53, 116
Surfactants, 53, 116
Suspensions, 56, 59, 116
Systeme Internationale, 5-6, 116

T

Tellurium, 94
Temperature, 8, 8t, 116
Thermonuclear reaction, 75-76
Titration, 85-86, 116
Trace metals, 91, 116
Transition metals, 90, 116
  chelates, 91, 92i
  trace, 91
Transmutation (nuclear), 76
Triple point of water, 8
Tyndall effect, 59, 116

U, V

Ultraviolet radiation, 15
Unit cells, 49, 50i
Valence electrons, 24, 26, 116
Van der Waals forces, 48, 115
Vapor pressure, 54, 116
Velocity, 6
Viscosity, 52-53, 116
Visible light, 14-15
Volume, 6

W, X, Y, Z

Water, 54-55
  calculating dissociation constant of, 84t
  ionization of, 84
Wavelength, 14
Weight, 63, 116
Wetting agents, 53
Work, 12, 116
X-ray diffraction, 49, 116

---

i refers to an illustration; t, to a table.